Management guidelines for Asian floodplain river fisheries

Part 1: A spatial, hierarchical and integrated strategy for adaptive co-management

FAO
RIES
CAL
PER

384/1

by
Daniel D. Hoggarth
Vicki J. Cowan
Ashley S. Halls
Mark Aeron-Thomas
MRAG Ltd, London, UK
J. Allistair McGregor
Centre for Development Studies, University of Bath, UK
Caroline A. Garaway
Renewable Resource Assessment Group, Imperial College, London, UK
A. Ian Payne
MRAG Ltd, London, UK
Robin L. Welcomme
Renewable Resource Assessment Group, Imperial College, London, UK

DEPARTMENT FOR
INTERNATIONAL
DEVELOPMENT
OF THE UNITED
KINGDOM

DFID

M_RAGLtd

Food
and
Agriculture
Organization
of
the
United
Nations

Rome, 1999

This document includes outputs from several research projects funded by the Department
for International Development (DFID) of the United Kingdom for the benefit of developing countries.
The views expressed are not necessarily those of DFID.

M-43
ISBN 92-5-104261-6

Preparation of this document

These guidelines summarise a series of studies, funded by the UK Department For International Development (DFID, previously known as ODA) on the management of Asian floodplain river fisheries. Between 1992 and 1997, DFID has funded four separate projects in this area. The research was initiated by a project led by Bath University's Centre for Development Studies (CDS), entitled 'Poverty, Equity and Sustainability in the Management of Inland Capture Fisheries in South and South-East Asia'. The theme was then taken up by three projects, led by MRAG Ltd, and funded by DFID's Fisheries Management Science Programme (FMSP): 'River and Floodplain Fisheries in the Ganges'; 'Fisheries Dynamics of Modified Floodplains in Southern Asia'; and 'An Evaluation of Floodplain Stock Enhancement'.

The Technical Paper is written in two parts. Part 1 outlines a practical strategy for the management of large floodplain rivers based on the experiences gained by these projects and on other literature from related research. Part 2 includes the more technical data derived from the research projects, on which the guidelines in Part 1 are based. The Part 1 guidelines are presented in a simple and user-friendly style to provide both policy makers and field officers with the tools they need to manage river fisheries, and the technical and institutional background to help make them work. The guidelines may also be used in the construction of courses for regional fisheries officers and related extension workers.

Funding for the preparation of this paper was provided by the UK Department For International Development. The Technical Paper was published through FAO to ensure the widest possible dissemination of its ideas. Over the years, the research theme has benefited from the active participation of a wide range of local collaborators, including staff from: the Bangladesh Agricultural University (BAU), Mymensingh, Bangladesh; the Bangladesh Centre for Advanced Studies (BCAS), Dhaka, Bangladesh; the Bangladesh Institute for Development Studies (BIDS), Dhaka, Bangladesh; the Central Research Institute for Fisheries (CRIFI), Jakarta and Palembang, Indonesia; the Environmental Laboratory of the University of Patna, India; the Coastal Resources Institute (CORIN) of the Prince of Songkla University, Hat Yai, Thailand; the Zoology Department of the University of Allahabad, India; the Zoology Department of the University of Garhwal, India. The many contributions of these collaborators to the ideas in this paper are warmly acknowledged. Special thanks are due to CRIFI's Dr Fuad Cholik, Dr Fatuchri Sukadi, Novenny Wahyudi, Dr Achmad Sarnita, Ondara, Agus Djoko Utomo and Zahri Nasution; to BAU's Dr M.A. Wahab; to MRAG's Prof. John Beddington and Kanailal Debnath; to CDS's Alex Kremer, Claire Hall, Dr Adrian Winnett and Prof. Chris Heady; to CORIN's Dr Somsak Boromthanarat and Dr Awae Masae; to the University of Patna's Dr R. Sinha; to BCAS's Dr Saleemul Huq; to Proshika's Rashed un Nabi; to DFID's Dr John Tarbit, Neil McPherson and Chris Price; and to FAO's Dr Jim Kapetsky.

MRAG Ltd is a leading international consulting firm, specializing in aquatic resources management, development, research and assessment, and information technology. The group has extensive experience of working in marine, freshwater, riverine and floodplain environments, and has worked in more than 60 countries for governments, private sector companies and international agencies. MRAG was formed in 1984, currently has a core staff of more than 20 professionals from a range of disciplines, and is located in the Imperial College campus, in the University of London. The group is associated with, and covenants funds to, the Marine Education and Conservation Trust, a charity which supports research and education.

Contact address:
MRAG Ltd
47 Prince's Gate
London SW7 2QA
UK

Hoggarth, D.D.; Cowan, V.J.; Halls, A.S.; Aeron-Thomas, M.; McGregor, J.A.; Garaway, C.A.; Payne, A.I.; Welcomme, R.L.
Management guidelines for Asian floodplain river fisheries. Part 1. A spatial, hierarchical and integrated strategy for adaptive co-management.
FAO Fisheries Technical Paper. No. 384/1. Rome, FAO. 1999. 63 p.

Abstract

This technical paper provides guidelines for an integrated management strategy for floodplain river fisheries. The paper is written in two separate volumes. Part 1 presents the guidelines in a 'user-friendly' format, to promote their uptake by fishery managers, policy makers and field officers. Recommendations are given both on the alternative technical tools which may be used to manage river fisheries, and on the institutional factors required for their success. The highly variable ecological and social characteristics of floodplain rivers demand locally-appropriate and adaptive solutions, rather than a single 'blueprint' approach. The recommended management strategy allocates responsibilities both hierarchically and spatially, and promotes the effective collaboration of government, communities and other stakeholders at appropriate levels.

The more technical Part 2 describes the underlying research work which provided much of the basis for these management guidelines. Investigations were made during four projects funded by the UK Department For International Development (DFID), in Bangladesh, India, Indonesia, Nepal and Thailand, between 1992 and 1997. Part 2 describes the floodplain river environments, the fish stocks and the fishing practices found at some of these study sites. Justification is given for a range of technical management tools for river fisheries, including the use of access controls and reserves, and the manipulation of water levels within flood control and irrigation schemes to give benefits to fishing as well as agriculture. Final chapters in Part 2 describe lessons learnt on the management of enhancement fisheries (e.g. based on fish stocking), and on the prospects and limitations of participatory management for these resources.

Distribution:

Inland fisheries: warm and cold waters
Directors of Fisheries
Fishery Regional Officers
Fisheries Department

Table of contents

List of Figures

Introduction

Attitudes to the management of natural resources are changing world-wide. These changes arise mainly from concerns about the state of the resources as they come under increasing pressure to satisfy a range of demands. Most important among these is the need for food, especially in tropical areas, which is forcing local populations to over exploit animals and plants. In addition, the sustainability of the living resources is threatened by impacts from other users by pollution and environmental modification. In general the capacity of present agricultural and industrial technologies to exploit and damage has far outstripped the capacity of societies to interpret, assimilate and control such changes. Efforts to do so show that present difficulties result from political, social and economic factors rather than from a lack of technological solutions. Concerns over these trends led to the convening of the United Nations Conference on the Environment and Development in 1992 and the acceptance of its Agenda 21. This highlighted the problems and, through Government commitment, provided a moral framework and guidelines for the sustainable use of natural resources. At the same time the Convention on Biodiversity was formulated and has now been accepted or acceded to by 176 Countries. The Convention is binding on its signatories and at present provides the only international legal framework for conservation and sustainable use.

Inevitably these developments have had their effect on fisheries. Most of the worlds fisheries are over-exploited or are about to become so. Nowhere is this more evident than in inland waters, especially rivers, which are almost without exception heavily or excessively fished. Added to this is the high degree of alternative use of water for industry, agriculture, power generation, urban supply and transport, all of which influence the amount of water in the system and the structure of the environment. As populations and levels of income rise so does the demand for water for these various uses and this commodity is now seen as limiting

development and human well being in many parts of the world. Unfortunately fisheries is one of the least valued of the various uses of water and in many areas institutions making decisions on the allocation of water do not even consider fish. As a result rivers are one of the most endangered ecosystems and their faunas are especially under threat of species extinction and population disturbance. To counteract these threats fisheries managers have to represent the interests of their sector in decision-making mechanisms at all levels. At the same time fisheries have to rationalise their own operations. One of the first steps in this process has been the adoption by countries of the Code of Conduct for Responsible Fisheries. This Code, adopted by the Food and Agriculture Organisation's Committee of Fisheries in 1995, furnishes voluntary guidelines on the organisation of fisheries at national level. It, together with the Convention on Biological Diversity, also emphasises the need for countries to regulate their aquatic environment for conservation of aquatic biodiversity and for sustainable fisheries.

Many of the ideas presented at these international fora are now being expressed at national level. As a result the objectives for management in many parts of the World are changing rapidly as policy makers adjust to the new vision of the resource. Frequently such changes exceeds the capacity of legislators to formulate new laws and it is probably better at present to retain flexibility through more generalised regulations until a more stable consensus emerges. At their meetings major discussions centre around mechanisms to ensure equity at national and international levels, and attempts to reconcile the conflict between conservation and use which lies at the heart of sustainable development. Furthermore, existing institutions and mechanisms have proved inadequate even for the management of food fisheries so new approaches to organising society for this purpose have to be sought. Three trends appear to be conditioning the direction of management at present.

The emergence of **new conservation oriented objectives** for natural resources management. This objective arises directly from the Convention on Biological Diversity and the Code of Conduct for Responsible Fishing and is more evident in temperate countries where abundant food supplies reduce dependence on river and lake fish. Nevertheless there is a growing trend to take these attitudes into account in attempting to make food fisheries more sustainable.

Changing patterns of use within the fishery. These arise mainly from demographic shifts throughout the world, principally human movements away from the rural to the urban. This has two effects. Firstly there is a need for fisheries for urban supply rather than satisfaction of local demand. Such fisheries tend to inflate the pressure on the resource by increasing prices and selecting for the larger fish in the assemblage. Secondly there is a global trend to reserve all or part of the resource for recreational rather than food fisheries. Usually the unit value of the recreational fishery is far higher than the food fishery. Where there is strong urban demand, recreational interests and pressure groups tend to displace the food fishery with drastic effects on food supplies and employment.

The emergence of **new philosophies for participatory management**. Two divergent trends are apparent here. On one hand, there is the suppression of national authority through economic integrating organisations. This is especially important in the case of fisheries where the basin approach is crucial for the protection of migratory or transboundary stocks. On the other hand, there is a trend for national powers to be devolved to local authorities within the country. This trend is further decentralised in the case of co-management systems where the fishermen's communities share responsibility for management. As the centralised nation state has been the preferred institution for management until recently these new approaches have yet to be developed and tested.

The study of large rivers and adequate understanding of how they function for fish and fisheries is fairly recent. For this reason the incorporation of these ideas into management of large floodplain rivers of the tropics is still developing. Further studies are needed particularly on the policy, social and economic aspects of the fisheries. There is also a need for research on the biology of individual systems to determine their state of exploitation and degree of modification by other users.

This paper presents the results of four research projects in South and South-East Asia. It also presents guidelines for management derived from these studies, adapted particularly to the large rivers of the region. These guidelines are intended to support the FAO Code of Conduct for Responsible fisheries and as such are aimed primarily at securing the sustainability of floodplain resources in order to secure a continuing supply of protein rich food. The Marine Resources Assessment Group carried out the work in association with several other overseas and UK institutes. Funding for this paper was provided by the UK Department For International Development. The Technical Paper was published through FAO to ensure a wide dissemination of its ideas.

The Technical Paper is in two parts:

Part 1. General ideas on the management of large floodplain rivers based on the experiences gained by the projects and on other literature from related research.

Part 2. Technical data derived directly from research on the selected South East Asian rivers.

The Technical Paper is aimed to provide fisheries administrators and scientists at policy making, executive and field levels with the tools they need to make decisions on the allocation of riverine resources and management of river fisheries. The material provided (especially in Part 1) may also be used in the construction of courses for extension workers.

Summary

This report provides guidelines for the management of fisheries in large floodplain rivers. Effective management of these complex resources requires holistic and multi-disciplinary approaches. The high variability of ecological and social characteristics between different rivers also demand locally-appropriate solutions. These guidelines thus attempt to show which questions need to be asked to find effective local solutions - there is no single 'right' answer which can be applied 'top-down'.

The guidelines are presented in a concise and simple format in Part 1. The underlying research which led to the guidelines is given in the more detailed Part 2. The Part 1 guidelines are written in five main sections - the 'why, what, who and how' of management, and a final summary showing how these components may be drawn together into an effective management approach. The guidelines recommend both a 'hierarchical' and a 'spatial' approach to management, based on the strong participation of both government, communities and other stakeholders at appropriate levels.

In the 'why' section, floodplain river systems are shown to be both highly *valuable* and highly *vulnerable*. Despite their high value, floodplain river habitats are now among the fastest disappearing of all ecological systems: effective management is urgently required. Floodplain rivers may be managed for many different objectives, but not all objectives may be achieved at the same time. Managers must, for example, choose between high employment combined with low profits, and low employment combined with a higher quality fish catch. To encourage local participation in management, the preferred objectives of fishing communities should be supported where possible.

The 'what' section describes the characteristics of floodplain fisheries resources, in terms of their environment, their fish and their exploitation by fishing communities. Multi-species, multi-gear floodplain fisheries have more complex interactions between the environment, the fish and the fishers, than any other type of fishery. This complexity may only be handled by an appropriate management approach, and the sharing of responsibilities between those most able to achieve them.

The floodplain environment comprises many different habitats and may provide many alternative livelihoods. With demands for irrigation water, power generation and flood control, floodplains are increasingly being modified on both a large scale and a small scale. Floodplain fish production, however, is dependent on the maintenance of the natural functioning of floodplain systems. Fisheries interests must be well represented in fora responsible for integrated catchment management.

Floodplains are inhabited by many different types of fish, including strongly migratory 'whitefish' and more locally-resident 'blackfish'. Whitefish and blackfish must be managed in spatial units appropriate to their distribution patterns: most whitefish will require a catchment focus, while blackfish may be managed more by villages for their own local benefits. The spatial relationships between waterbodies and the communities near to them will determine who may be able to manage effectively in each locality.

The high species and habitat diversity of floodplains is reflected in the complexity of the fishery. Floodplain fishing communities often comprise a complex network of 'stakeholders', with leaseholders, middle-men and fishers at various levels of authority and dependency. Many different types of fishing gears are also used, from simple hooks and traps up to more elaborate, expensive and effective structures such as barrier traps and fish drives. 'Hoovering gears', such as dewatering, poison, electric fishing and fish drives attempt to catch all the fish in dry season waterbodies, and must be limited to ensure the survival of blackfish as they prepare for spawning with the new flood. Barrier gears must also be limited to ensure the access of whitefish to their spawning grounds.

Floodplain river fisheries need local solutions - there is no single 'right' answer which may be applied 'top-down'

The next section considers *'who'* should be involved in the management of floodplain fisheries. The first step is to identify who has an interest or *stake* in the fishery. A description of stakeholders and the nature of their interest is vital information as management is essentially about managing the people who exploit the fishery. Understanding the relationships between stakeholder groups is also important, for example whether they co-operate or conflict with each other or whether the groups are dependent in some way (financial, political etc.) on each other. Awareness of the relationships that stakeholders have with the fishery and with each other will guide the development and implementation of management plans. For example, activities that will increase conflict, and so undermine the chances of management success, can be avoided or planned for (i.e. the activity can be gradually introduced alongside measures designed to improve relationships and trust between the groups in conflict). Exclusion of any one stakeholder group from the process can seriously undermine management, rules can be ignored and challenged as people feel they have had no part in their development. Awareness, consultation and the full participation of stakeholder groups will usually increase the perceived legitimacy of management and the likelihood of its success.

An indicative list of the groups who will have a stake in floodplain fisheries management is presented alongside a brief description of their likely interests in the fishery. The word 'community' is used in this listing, with the reservation that it implies a cohesive group of people with similar objectives. In reality, 'floodplain communities' are typically very mixed, often with significant social, cultural and financial differences. A stakeholder analysis of a specific community would disaggregate the community into smaller groups with similar interests. The process of developing management plans could then take account of these differences between groups.

Successful management requires that stakeholders take responsibility for a range of roles. This document identifies and discusses 18 individual roles, as follows: establish management objectives;

ensure international responsibilities are taken into account; ensure the environment is protected; assess the fishery; provide technical guidance (knowledge/expertise); conduct research (pure and applied); provide a catchment perspective for management; develop management plans; set rules for fishing (i.e. who can fish, which species, where, when and how); set rules for institutional support of fisheries management; develop appropriate legislation to support fisheries management; provide mechanisms for conflict resolution; co-ordination; communication; training/extension; monitoring; enforcement; and, funding of fisheries management. The list is long, but not exhaustive. It is intended to guide managers as they develop their own priorities for successfully managing their fishery.

The question of who should take responsibility for which role is not always straightforward. The decision requires an assessment of the 'capacity' of each of the stakeholder groups. Four categories of capacity are identified: resources, trained members, rights and motivation. Lack of resources often undermines fisheries management. Resources may include staff in a Fisheries Department or members of a co-operative management group as well financial resources. The costs associated with some management interventions, such as stocking and habitat rehabilitation, can be very high and shortfalls in a budget may be more of a constraint than technical or social aspects of the intervention.

The second category is trained members. Fisheries management requires a range of skills and stakeholder groups must have people who are experienced or well trained in the areas for which they have responsibility. As can be seen from the list of roles identified here, skills in technical assessment of a fishery through to conflict resolution are needed for successful management. Often these skills are not present or will need to be improved by appropriate training.

The third category of capacity is rights, this covers recognised roles, responsibilities and the right to manage. With so many stakeholders involved in floodplain fisheries and the long list of roles, it is very important that all groups involved understand and agrees who is doing what. Conflict often occurs in fisheries when there is some confusion or disagreement between groups over responsibilities: this can not be ignored as such conflict undermines management. The recognition of different groups' rights to manage often needs to be formal. This can be through a formal letter of agreement between the floodplain stakeholders and the relevant government department or even included as part of national or local legislation. This provides the stakeholders with the authority to manage including the power to exclude groups who are not part of the agreement. This is a very powerful and important part of a stakeholder group's capacity to manage.

The final category is motivation. Fisheries management requires active involvement of all stakeholders to the process of developing, implementing and refining plans for management. Involvement takes time and sometimes money, therefore incentives and disincentives need to be understood. Stakeholder analysis will help identify the reasons why people will sometimes prefer not to be part of the management process: plans need to be made to maximise incentives and minimise disincentives. Incentives can range from adequate salaries and opportunities of promotion for staff in government organisations, through to tangible benefits such as larger fish and improved social status for members of stakeholder groups based on the floodplain.

The final part of the 'who should manage' section proposes a match of stakeholders to the necessary management roles within a hierarchical system of management units. Three very broad groups of stakeholders are identified for this exercise: government departments, floodplain communities and intermediary organisations (i.e. projects, NGO's, aid agencies etc). As with the listing of roles, this match is presented as a guide to managers, providing an overview and illustrating the process rather than giving a definitive answer for all floodplain fisheries. The main outcome of this is

that all three groups have important roles to play in both catchment and local management areas. However, responsibilities will vary between groups depending on which level of management area is under discussion. Governments will have to take the lead in catchment level management as they have the greatest capacity to coordinate and manage fisheries on this scale. In contrast, floodplain communities are often in a better position to take responsibility for managing the local areas within a supportive framework provided by government. Intermediary organisations may play a facilitating role, supporting both governments and communities in the clarification of roles, improving capacity and developing management plans.

Finally then, the '*how*' section presents guidelines for the sub-division of floodplain rivers into a hierarchy of spatially defined 'management units', and for their adaptive management using locally appropriate tools. Responsibility for the management of a river fishery should be divided between a 'catchment management authority' and a number of other co-management partnerships, each managing a local sub-unit. Such partnerships may be either village-based or district-based, depending on the size and types of waterbodies involved and on the traditional and formal activities of existing institutions.

Fishery management units should be selected to achieve the maximum overlap between the range of authority of the management group (e.g. a village boundary) and the distribution range of a fish stock. The managers of Catchment Management Areas (CMAs) should be responsible for (1) managing migratory whitefish stocks for the overall benefit of the catchment's fishers, (2) co-ordinating management activities in the smaller village units, and (3) representing fishery interests in sectoral talks on integrated catchment management. The smallest Village Management Areas (VMAs) provide the strongest management opportunities for local blackfish stocks, where fishing communities have traditional control over local waterbodies, within areas small enough to manage effectively. Other Intermediate Management Areas (IMAs) may also be identified in between CMAs and VMAs. In both IMAs and VMAs, traditional institutions and government administrative systems should be built upon where available to take advantage of existing management skills, local knowledge and systems of authority.

Guidelines are given for the strategic assessment of the different types of management units, and for the preparation of management plans. Such plans may include a mixture of different tools for ensuring sustainability, raising revenues, and ensuring a fair distribution of benefits. The appropriate management plan for each fishery unit depends on its local environmental and social conditions, its management objectives, and its current state in comparison with those objectives. Menus of different tools are proposed for each type of management unit, and the objectives, advantages and disadvantages of each tool are described.

Due to the competitive interactions between fishing gears, it is important to recognise that technical management approaches such as gear closures, closed seasons, and mesh/fish size limits always have benefits for some gears, and losses for others. Barrier trap regulations and habitat restorations are particularly recommended for CMAs to maintain the long-distance migration pathways of whitefish. Restrictive leasing of waterbodies combined with reserves and other technical tools are recommended for VMAs to ensure the dry season survival of local blackfish stocks and the access of whitefish into community waters. Simple management tools are recommended for larger floodplain lakes shared between surrounding villages, to control overall fishing levels, encourage co-operation, and reduce conflicts. Tools for floodplains impounded by flood control or irrigation schemes are also described, mainly involving the operation of sluice gates for the joint benefits of agriculture and fisheries.

Due to the high variability between floodplain systems, the local impacts of such different management tools are impossible to predict in advance. An *adaptive management* approach is therefore recommended, with monitoring programmes designed to ensure the local

effectiveness of the chosen management strategies. The participation of fishers in these monitoring programmes is encouraged, and the funding required should be raised from the fishery wherever possible. The results of the monitoring programme should be widely disseminated, and should show clearly whether or not the community's objectives for the fishery are being achieved.

The information used to monitor the fishery will depend on the objectives chosen for it and may include both ecological and socio-economic data. Changes in the *inputs* to the fishery should be monitored in order to understand any changes in the monitored *outputs*. Monitoring changes in the wider environment or fishing practices, for example, may help to explain unexpected changes in fish catches. Since environmental conditions change naturally from year to year, managers must be prepared to examine long-term trends in their data. 'Process monitoring' of the co-management partnership may also be vital, especially in the early stages, to ensure that conflicts of interests or other factors do not prevent effective collaboration.

The final section of the guidelines summarises the necessary steps required for successful management. The steps are divided into those which should be taken by (1) national-level policy makers, (2) catchment managers (including any co-ordinating fora for multi-CMA rivers), and (3) managers of the individual CMA, VMA and IMA fishery units. Though the ongoing adaptive management process is relatively simple, it can only be successfully implemented when all the necessary steps have been implemented at higher levels. Most importantly, decentralised management can not proceed effectively until the management rights of local people and agencies have been recognised and clearly stated in the legislation. With appropriate hierarchical and spatial sharing of management responsibilities between government and local communities, and with effective participation and communication, the high value of floodplain river fisheries may be retained by these approaches.

Part 1 - Management Guidelines

Part 1 of this paper provides guidelines for the management of large floodplain rivers. Effective management of these complex resources requires holistic and multi-disciplinary approaches. In addition, the high variability of ecological and social characteristics between different rivers demand locally-appropriate solutions. These guidelines attempt to show which questions need to be asked to find effective local solutions - there is no single 'right' answer which may be applied 'top-down'.

The guidelines are presented in a concise and simple format, leaving the underlying research to the more detailed Part 2. The concepts are presented in four main sections - the 'why, what, who and how' of management. A final summary section shows how these components may be drawn together into an effective management approach. The 'why' section demonstrates the high value of floodplain fisheries and their urgent need for management. The 'what' section describes the characteristics of floodplain fisheries resources, in terms of their environment, their fish and their exploitation by fishing communities. The next section considers *who* has an interest or *stake* in the management of the fishery, and *who* has the capacity and ability to fill the various roles required. Finally, the 'how' section presents guidelines for the sub-division of floodplain rivers into spatially defined 'management units', and for their adaptive management using locally appropriate tools. The guidelines recommend both a 'hierarchical' and 'spatial' approach to management, based on the strong participation of both government, communities and other stakeholders at appropriate levels. This approach represents the logical outcome of the various studies described in Part 2, but now requires field testing and validation. The authors would thus welcome any comments on these ideas.

1 Why Manage?

1.1 Why manage floodplain rivers?

Floodplain river systems are both highly valuable and highly vulnerable. Both of these characteristics are partly due to the impacts of external factors on the resource. Water flows from upstream bring both beneficial nutrients and potentially damaging pollutants. They are also partly due to the extensive and variable nature of the floodplain environment: this provides many opportunities for natural resource use, but also stimulates over-use and destruction when different users compete for access.

The high values of floodplain river systems are due to:

- their high biological productivity (and high potential value of exploitable resources),
- their high resilience to heavy exploitation levels and climate changes,
- their high biodiversity, and
- their multiple alternative livelihood opportunities.

Their high vulnerability is due to:

- the often conflicting demands of different sectors (e.g. fisheries, agriculture, transport, forestry, water abstraction, water drainage, housing, industry....), and
- negative impacts from upstream sources (e.g. pollution, deforestation...).

Inland waters of the Asian region (including rivers, lakes and reservoirs) are more heavily exploited than in either Africa or South America, and provide more than half of the world's production from inland capture fisheries (52.3% of the world catch of 6.5 million tonnes in 1990, according to FAO).

Despite their high values, floodplain river habitats are now among the fastest disappearing of all ecological systems. Outside Asia, it has been estimated that 77% of the 139 largest river systems in North America, Europe and the former

Floodplain river systems are both highly valuable and highly vulnerable

Despite their high values, floodplain river habitats are now among the fastest disappearing of all ecological systems

Soviet Union are strongly or moderately affected by interventions associated with navigation, reservoir operations, inter-basin diversions and irrigation. These modifications have resulted in the loss of many fisheries particularly for migratory species in both the rivers themselves and in the lakes and seas they feed. In other cases they have necessitated intensive efforts to retain valuable species, such as the sturgeon and the salmon, through artificial breeding and stocking programmes. There are strong pressures for similar modifications in large rivers everywhere. The fisheries sector is frequently viewed as economically unimportant when compared with the financially powerful electricity generation lobby, with the needs for supply to the urban, industrial and agricultural sectors. For this reason it tends to be overlooked when allocating water or issuing permits for public works that involve massive alteration to the aquatic ecosystem. Such abuses can only be avoided if fisheries interests are strongly represented in those fora responsible for such decisions.

Floodplain rivers thus urgently require effective, integrated management to ensure that their potential values to fisheries and other sectors are maintained. This urgency has already been recognised by the Ramsar Convention, under which a few particularly valuable wetland environments are protected as nature reserves. Much more needs to be done for the vast majority of river catchments which lie outside Ramsar sites.

1.2 Management objectives

Well-managed fisheries may be highly productive, and may serve many different objectives. As shown below, different objectives will appeal to different levels of society. Unfortunately, not all of these objectives can be achieved at the same time. Managers must thus attempt to satisfy as many objectives as possible, and must recognise that their goals for the fishery, such as maintaining biodiversity or raising revenues, may not all be shared by fishing communities.

These guidelines show how different objectives may be achieved by *collaborative* management. As will be seen, successful management requires appropriate decision-making and effective contributions from each of the above management levels.

Likely selection of alternative management objectives by different levels of society

	International Policy Makers	National Policy Makers	Regional Fishery Managers	Local Communities
General Objective				
Sustainability (of following)	✔	✔	✔	✔
Ecological Objectives				
Biodiversity	✔	✔		
Conservation	✔	✔	✔	
Primary Use Objectives				
Food / nutrition	✔	✔	✔	✔
Ornamental fish		✔	✔	
Sport fishing		✔	✔	
Social Objectives				
Income to fishers				✔
Equity / benefit distribution		✔	✔	✔
Employment		✔	✔	✔
Poverty reduction	✔	✔	✔	
Conflict reduction	✔	✔	✔	✔
Government Objectives				
Revenue to government		✔	✔	
Contribution to GDP		✔		
Export income		✔		

2 What to Manage?

2.1 What makes floodplain river fisheries special?

All fisheries depend on an interaction between the environment, the fish which depend on that environment, and the fishers who catch the fish. As illustrated below, the complexity of each of these factors is at a maximum for floodplain fisheries resources:

2.2 Environment

2.2.1 River floodplains and the wider catchment

Floodplains are the most highly productive part of any river system. Their productivity derives from both the inputs of nutrients from upstream and the seasonal recycling of plants and

Multi-species, multi-gear floodplain fisheries have more complex interactions between the environment, the fish and the fishers, than any other type of fishery

Resource Component	Simple fishery (e.g. lake or marine trawl fishery)	Floodplain River Fishery
Environment	Stable over time Single habitat Resource mainly used for fishing	Seasonal fluctuations within year Variable flooding between years Many habitats Habitats vary between localities Strong competition for resource use
Fish	Single / few species	Multiple species Variable behaviours and requirements
Fishing	Single gear type Commercial / capital intensive Similar fishing communities Few central landing centres	Numerous gear types Artisanal / labour intensive Different fishing communities Many dispersed landing centres

This complexity may partly explain why relatively little attention has been given to floodplain river capture fisheries, compared to marine fisheries. A further factor may be the importance of *local* conditions on the effectiveness of different rules (including government regulations and traditional rules). This dependence prevents the use of a single, standard management approach for all floodplain resources. Though floodplain fisheries *are* complex, this complexity *is* manageable, given the right management approach, and a clear sharing of responsibilities.

The rest of this section describes the particular characteristics of floodplain fisheries, in three categories: environment, fish and fishing. Sections 3 and 4 further develop the implications of these characteristics on *who* may best manage floodplain fisheries, and *how*.

animals which occurs with each 'flood pulse'. Though the main river channels supply the floodplain with nutrients, they are relatively unproductive themselves, due to their strong currents and shifting substrates. They may also bring down any negative impacts of poor management from upstream: both the *quality* and the *quantity* of water in rivers is vital for maintaining productivity. Reductions in water flows may be caused by diversion of water into upstream irrigation schemes. Dangerously high flooding and dry season water shortages may both be caused by deforestation, when water runs more quickly off logged hillsides. In large rivers, such impacts may flow across international borders.

River catchments, and the highly productive floodplain systems in particular, may thus provide many alternative livelihood opportunities. With careful planning, such activities may be compatible. There may also, however, be negative interactions, particularly where sectors are unaware

Though floodplain fisheries are complex, this complexity can be made more manageable by the right management approach and a clear sharing of responsibilities

of each others needs. With multiple resource use, it may often be difficult to determine which sector is actually causing a given impact.

2.2.2 High spatial and seasonal variability

River habitats change gradually from the fast-flowing upland streams to the slow, meandering lowland rivers. Habitat diversity is highest in the floodplain sections, with flooded grasslands, flooded forests, small and large river channels, and permanent and temporary lakes and pools. Each of these habitats is used by different fish species for their essential life processes, such as spawning and feeding. Key habitats are normally accessible through floodplain channels: these may need to be maintained in highly exploited or modified river systems. The mixture of habitats varies significantly between localities and determines the types of management measures likely to be of use. Where maintained, the complexity of the resource may support extremely high fish biodiversity.

The floodplain environment also varies seasonally, both within the year, and between different years. The annual cycle divides the year into periods of high fish productivity during the flood season, and relative inactivity and hardship during the dry season. Variability in the size and duration of these seasons affects the productivity of the floodplain and the effectiveness and profitability of the fishery. Variability in the timing of the seasons prevents the use of rigidly-timed management frameworks.

2.2.3 Floodplain modification

Floodplains are increasingly being modified on both a large scale and a small scale. Governments are building dams, impoundments and polders to generate electricity and control flooding. Local communities are reclaiming floodplain land for farming, and digging fish pits to catch fish. River channels are becoming blocked by siltation. Though sometimes beneficial to other sectors, such changes can have

significant impacts on fisheries productivity. Though the impact of individual small-scale modifications may be minor, their cumulative effect may be large.

> ### *Management implications of the floodplain river environment*
>
> - Managers of fisheries and other resources must discuss their impacts on each other
> - The impacts of floodplain modifications must be investigated and managed at both catchment and local levels
> - Variability in habitats between localities necessitates local involvement in management
> - Uncertainty in hydrological regimes necessitates use of flexible management systems
> - Quantity and quality of flood water must be maintained for high fish productivity
> - Diversity of floodplain habitats must be maintained for high fish biodiversity
> - River channels must be maintained for fish migrations and access to spawning grounds

2.3 Fish

2.3.1 Types of floodplain river fish

Tropical floodplain river fish stocks may comprise over 200 different species of fish. Around 30 different fish species are commonly caught by floodplain fishers in any one locality. Each species clearly can not be managed individually, and it is usually necessary to group species into management units, or 'guilds'. For this purpose, floodplain river fish may be categorised in one or more of the following ways:

- **Migration patterns:** local ('blackfish') and long-distance ('whitefish')
- **Feeding:** predators, herbivores, others
- **Taxonomic groups:** carps, catfish, perches, snakeheads etc.
- **Sizes:** large, medium and small
- **Values:** high, medium and low

From a management perspective, the first two categories (migration and feeding) are the most important. Fish migrate to find the best conditions for breeding, feeding and survival in different parts of the river system. Some 'whitefish' migrate thousands of kilometres up and down rivers, while other 'blackfish' may spend most of their lives in a single waterbody (see Figure 2.1). Blackfish species are able to tolerate the de-oxygenated conditions of the dry season in floodplain waterbodies while whitefish usually return to the main river or large lakes to survive. In Asian rivers, blackfish include species such as the snakeheads and the climbing perch (*Anabas testudineus*), while whitefish include many large carps and riverine catfish in addition to the valuable giant prawn (*Macrobrachium rosenbergii*). The alternative migration patterns of blackfish and whitefish determine whether they are vulnerable to many different fishing communities across the river catchment or to only one or a few local ones. To be effective, fisheries managers must regulate the activities of fishing communities across the full range of a species' distribution. Whitefish species must therefore be managed in much larger '*management units*' than blackfish.

The combination of migration patterns and feeding behaviours determine which species are caught by which fishing gears, and at which times. Strongly migratory whitefish, for example are caught by barrier traps; predatory fish are caught by baited hooks; air-breathing blackfish are caught by fish drives in dry season floodplain pools. Multi-species floodplain fish stocks have many 'interactions' with the multi-gear fishery: every type of fishing gear always catches several types of fish, and every fish species is always caught by more than one fishing gear. Fish behaviours thus determine which species will be affected by management regulations on certain gears or certain seasons. The strong interactions between gears and fish mean that no single gear or fish species should be managed independently of the overall fishery.

2.3.2 Impact of fishing

Asian river fisheries are highly productive, with average catches of around 100kg per hectare of floodplain. Surprisingly, this overall catch rate is not strongly affected by the amount of fishing. Though heavy

Migratory whitefish must be managed in much larger management units than local blackfish

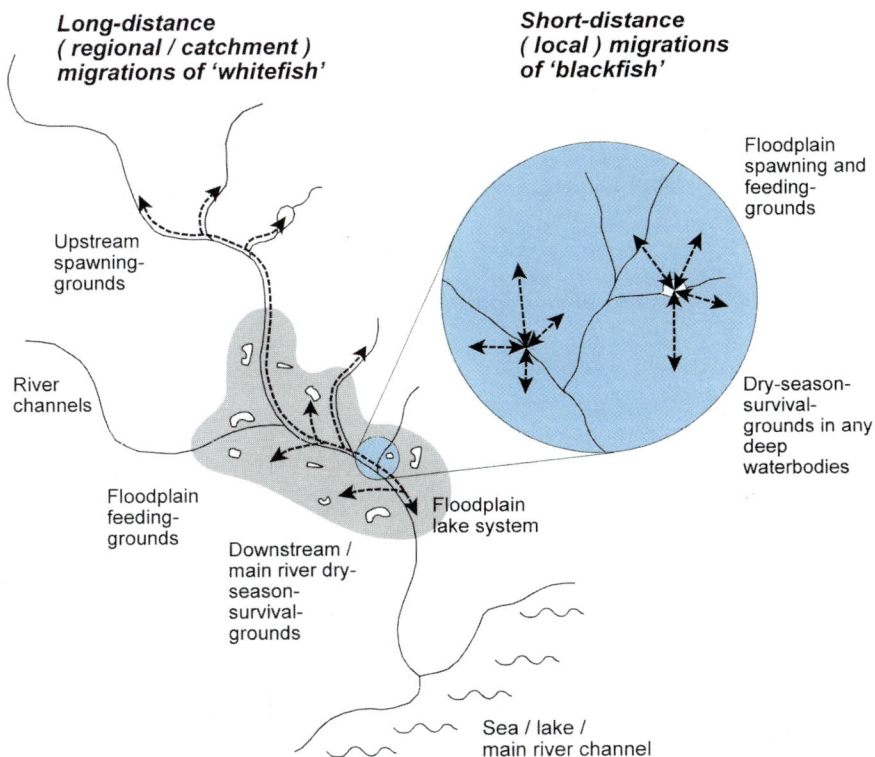

Long-distance (regional / catchment) migrations of 'whitefish'

Short-distance (local) migrations of 'blackfish'

Upstream spawning-grounds

River channels

Floodplain feeding-grounds

Downstream / main river dry-season-survival-grounds

Floodplain lake system

Floodplain spawning and feeding-grounds

Dry-season-survival-grounds in any deep waterbodies

Sea / lake / main river channel

Figure 2.1 Alternative migration patterns of riverine 'whitefish' and 'blackfish'

fishing may over-exploit certain species, these may be replaced by other members of the multi-species stock (see top two graphs in Figure 2.2). As long as stocks are not completely wiped out (e.g. by fishing with poison), multi-species floodplain river fisheries can thus maintain high catch rates even with extraordinarily high levels of fishing effort.

Heavy fishing of floodplain fish stocks thus mainly affects the *species* of fish caught, not the total weight of the catch. Since fishers tend to exploit the most valuable fish species first, these are usually the first to decline. Though catches may remain high in a heavily exploited fishery, their value may eventually decline to the point where they are less than the costs invested in fishing for them (see bottom graph in Figure 2.2). Managers must thus choose whether to allow heavy fishing for very little profit (e.g. where the objective is to generate employment or provide nutrition to poor people) , or to restrain the amount of fishing to improve the types of fish caught and the overall value of the catch.

2.3.3 Impact of other sectors

As discussed in other sections, the activities of other resource-users in a river catchment may also affect the productivity of the fishery. The impacts of pollution, and losses of spawning habitats, for example, may be far more significant than those from fishing.

Such impacts may be hard to detect or even investigate, but their existence calls for active participation in inter-sectoral talks on the use of the shared river resource.

2.4 Fishing

2.4.1 Local control of waterbodies

Fishing communities are usually distributed throughout the floodplain river system. The fishing opportunities of each community depend on the types and sizes of waterbodies to which they have access. In some localities, small floodplain waterbodies may be traditionally associated with villages, or 'owned' by them. Other waterbodies, particularly larger lakes, main river channels and open floodplains, may be less easy for a single community to control, and may be openly accessible to fishers from many surrounding communities. The spatial relationships between riverine waterbodies and the communities near to them thus determine who may be able to manage effectively. Where traditional management exists, this should be encouraged and built upon, to take advantage of existing skills.

2.4.2 Floodplain river fishers and access to fishing

Floodplain rivers are fished by many different types of people. Fishers may be full-time professionals, often working as

---

Management implications of floodplain fish stocks

- Migration patterns of '*blackfish*' and '*whitefish*' determine fishery '*management units*'
- Due to fish movements, management units can rarely be managed in isolation (villages may participate in management of their local waters, but higher authorities must facilitate communication between adjacent management units)
- Multiple interactions between fish and gears mean that no fish species or gear type should be managed on its own
- Since floodplain fishing depletes the large and valuable fish species first, managers must choose between high employment in the fishery or high value of the species caught
- Due to the impacts of external factors on fish stocks, fishery managers must participate in talks on '*integrated resource management*

groups with expensive fishing gear, or only part-time, perhaps working on their own with more simple gear. Part-time fishers often alternate seasonally between fishing and other occupations, such as agricultural labouring. 'Subsistence' fishers are usually dependent on fishing to provide their daily meals. In poor, heavily-populated countries, where the costs of some types of fishing may be very low and alternative labour opportunities very limited, the numbers of fishers can rise to exceptionally high levels.

These different types of fishers rarely operate independently of each other. Floodplain fisheries often have complex arrangements (traditional or formal) for controlling access of fishers to different waterbodies. Poorer fishers are often unable to pay for expensive fishing fees up-front, and must rely on credit in some form. This may give rise to sub-

licensing of individual gears or fishers within sub-divisions of waterbodies. The fishing community may then be a complex network of 'stakeholders', with leaseholders, middle-men and fishers at various levels of authority and dependency. Where they exist, such community networks may be valuable instruments for improving control of the fishery.

2.4.3 Floodplain river fishing gears

The diverse habitats and fish species of floodplain rivers are reflected by an equally diverse range of fishing gears and operations. Each river may be fished by twenty or more gears, which may be categorised into one of the following four types (see Figure 2.3):

- *Set-and-wait* gears are set and hauled after a few hours, and catch fish

Hoovering gears must be managed to ensure the dry season survival of blackfish

Barrier gears must be managed to ensure the access of whitefish to their spawning grounds

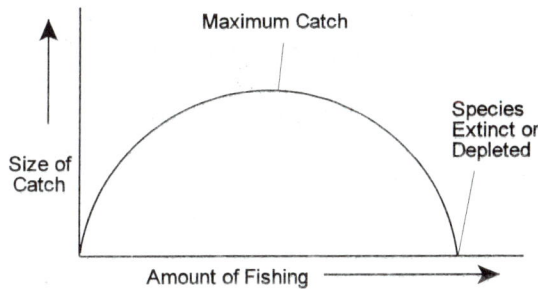

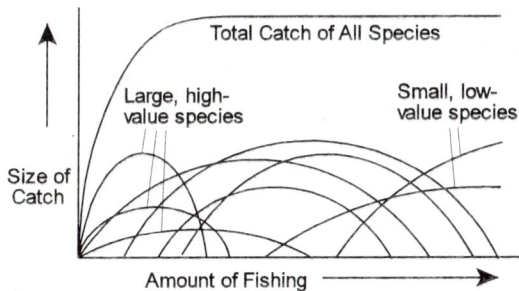

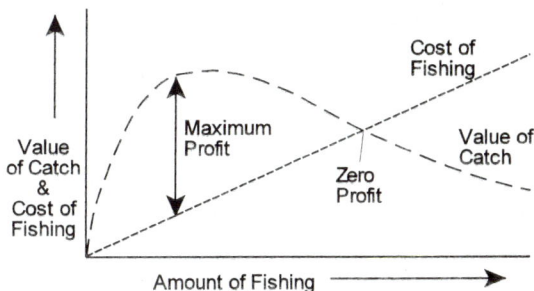

Figure 2.2 Simple relationships between fishing, catches and profit for a multi-species river fishery

The spatial
relationships
between
waterbodies and
the communities
near to them
determine who
will be able to
manage
effectively

when they are feeding or moving around the floodplain (e.g. gill nets, portable traps and baited hooks).

- *Chasing* gears involve more active pursuit of fish by fishers, sometimes even in open floodplain waters (e.g. drag nets, push nets, some types of seine nets and spears).

- *Barrier* gears are set more permanently than set-and-wait gears to catch whitefish during their seasonal migrations. Barrier gears may be set both in the main river (where it is narrow enough), or in the secondary channels where fish migrate off the floodplain when waters fall. They may also be complete barriers (e.g. suspended trawl nets, bamboo barricades) or only partial barriers, which do not span the full width of the channel (e.g. lift nets, fyke nets etc.).

- *Hoovering* gears are used in the dry season to catch blackfish stranded in pools and river channels (e.g. fish drives, electric fishing, and - most dangerously - poison and dewatering).

This wide variety of fishing gears enables the capture of the many different fish species in all the different floodplain habitats, and in the various

seasons of the year. Set-and-wait gears and chasing gears may catch fish at any time, but are usually labour intensive and fairly inefficient. Barrier and hoovering gears, in contrast, are very much more efficient as they catch fish at those times when they are highly concentrated in specific habitats (barriers in channels, and hoovers in dry season waterbodies – see Figure 2.4). Access to such habitats will usually be in strong demand and thus require some management to prevent conflicts.

The relative numbers of each type of gear used in a fishery determines the distribution of catches between the members of the fishing community. Many of the fishing gears will compete for the same fish species, so that the excessive use of one gear, say in the flood season, will decrease the catches of a second gear used either at the same time or later in the dry season. Owners of large barrier and hoover gears may thus wish to prevent the excessive use of gill nets during the flood season to maximise their own catches during the drawdown. Where such large gears are expensive, their profitability may indeed only be assured by limiting the earlier fishing of other small gears. A fishery with many set-and-wait and/or chasing gears will have a wide distribution of

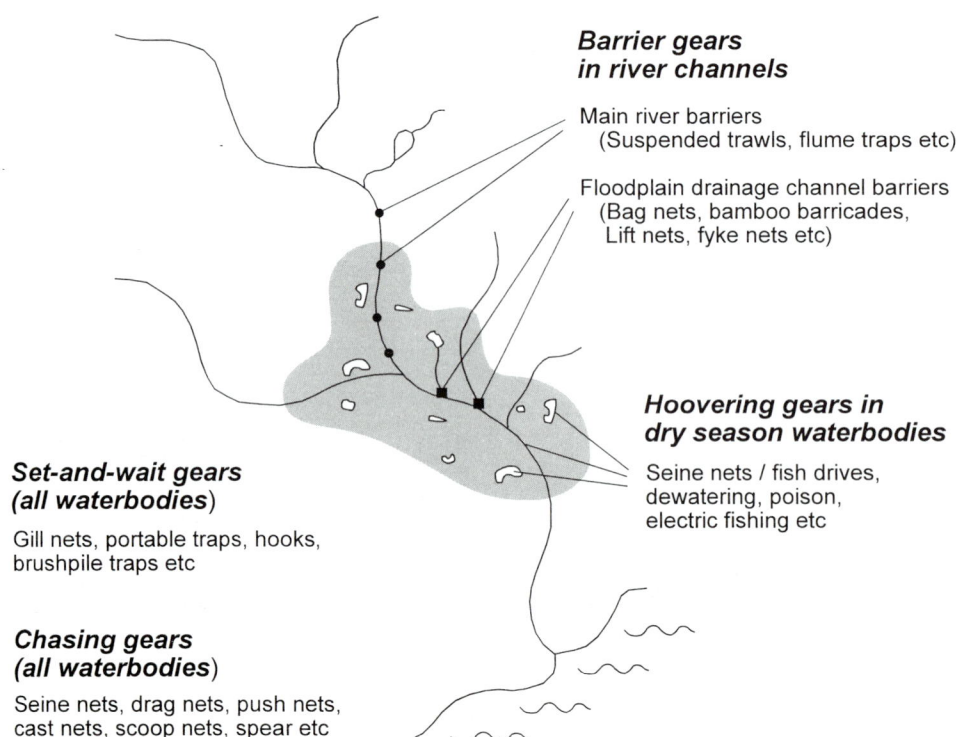

**Barrier gears
in river channels**

Main river barriers
(Suspended trawls, flume traps etc)

Floodplain drainage channel barriers
(Bag nets, bamboo barricades,
Lift nets, fyke nets etc)

**Hoovering gears in
dry season waterbodies**

Seine nets / fish drives,
dewatering, poison,
electric fishing etc

**Set-and-wait gears
(all waterbodies)**

Gill nets, portable traps, hooks,
brushpile traps etc

**Chasing gears
(all waterbodies)**

Seine nets, drag nets, push nets,
cast nets, scoop nets, spear etc

**Figure 2.3 Types of
floodplain river
fishing gears**

fish catch among the community. In contrast, one with only barrier and hoover gears will usually have less operators and a narrower distribution of catches. Most fisheries involve a balanced mixture of both types of gear.

The high effectiveness of barrier and hoovering gears makes them particularly threatening for the long-term sustainability of fish stocks. Managers must ensure that barrier gears do not prevent the upstream spawning migrations of whitefish, and that hoovering gears do not catch *all* the blackfish during the dry season. The use of barriers during the rising water seasons, or of poisons, electric fishing or dewatering during the dry season are especially dangerous.

Management implications of floodplain communities and fishing

- Fishing communities with existing management systems for local waterbodies provide strong opportunities for management

- River fisheries should be managed as a whole, not as individual gears (management regulations affect the *distribution* of the catch between gears, more than its total size)

- Stock ownership regimes may be encouraged in appropriate waterbodies, to increase local incentives for long-term conservation

- Hoovering gears must be managed to protect blackfish (ensure dry season survival)

- Barrier gears must be managed to protect whitefish (ensure access to spawning grounds)

- Floodplain *livelihoods* should be managed: fishing is *not* the only opportunity (talk with managers of other sectors in catchment; provide retraining, relocation etc.)

Floodplain fishing communities often comprise a complex network of stakeholders, with leaseholders, middle-men and fishers at various levels of authority and dependency

---

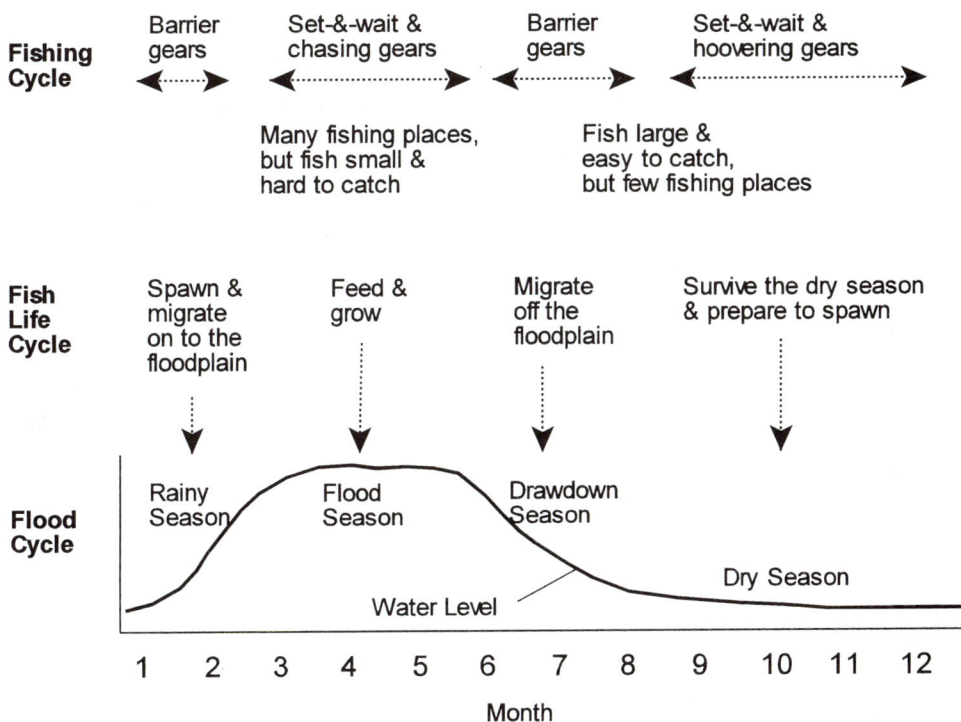

Figure 2.4 The relationships between the seasonal cycles of fishing, fish biology and flooding

3 Who Should Manage?

3.1 Introduction - sharing roles and responsibilities

The preceding sections have emphasised the high value of floodplain fisheries, their complexity and diversity, and their vulnerability to both over fishing and degradation from other sources. While there is clearly a need for good management to maintain productivity, this may seem an overwhelming task for these complicated systems. This section attempts to resolve this problem, by showing how the various management roles and responsibilities may be *shared* between a range of collaborators. This sharing may take place both *hierarchically*, as in a 'co-management' relationship between government, local people and other organisations, and *spatially*, between different geographic sub-units of the fishery. The remainder of this introduction briefly describes these two types of sharing, while the rest of Section 3 explores who may take on the various roles and responsibilities.

3.1.1 Hierarchical sharing: co-management

Co-management has been described as a 'partnership arrangement using the capacities and interests of the local fishers and the community, complemented by the ability of government to provide enabling legislation, enforcement and conflict resolution, and other assistance'. Co-management requires increased emphasis on communication and the use of flexible approaches to manage successfully, and is seen as a solution to some of the problems experienced by the 'top-down' use of standard technical solutions.

The involvement of both government and the community is recommended for both technical and social reasons. Locally appropriate management solutions may best be identified by combining the community's knowledge of local resources with the government's more conceptual knowledge of the dynamics of the whole fishery. One of the greatest incentives for fishers to follow rules and observe agreed practices is for them to have some involvement or responsibility in the management of the fishery.

As illustrated in Figure 3.1., co-management may comprise various levels of sharing, anywhere along a line between fully centralised government management, and totally independent 'bottom-up' self-management by communities. The position adopted along the line should reflect the nature and scale of the management problems at hand and the abilities and capacity of each of the different collaborators.

While co-management approaches are recommended here for floodplain fisheries, it must also be recognised that this approach will not always be possible. The movement of fish around the river means that management will always be more complicated for fisheries than for other non-mobile resources such as trees. In addition, conflicts of interests between different social groups, or the self-interests of even a small number of locally powerful community members may prevent any partnerships from working effectively. Certain environmental and social situations will thus be more appropriate for co-management than others, as illustrated later in this section. While some parts of each river should be expected to present major management challenges, other parts may often provide good opportunities. These latter parts are where co-management activities should begin.

Figure 3.1 Co-management as a range of partnership arrangements between central government and local communities.

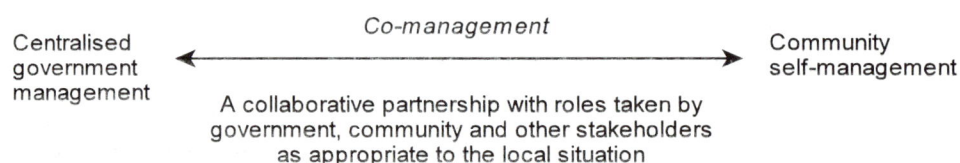

Centralised government management

Co-management

$\longleftrightarrow$

Community self-management

A collaborative partnership with roles taken by government, community and other stakeholders as appropriate to the local situation

3.1.2 Spatial sharing: fishery management units

As described in Section 2, floodplain river fisheries are only part of the wider river environment. Interactions of the fishery with other sectors, such as agriculture, will usually need to be managed at a catchment-wide level. *Whitefish* species which migrate around the full river system must also be managed at this level. Such management activities are best handled by government, with their regional perspective and authority, and access to the departments responsible for other sectors. Local communities will have relatively minor roles at this level.

In contrast, local communities may play strong roles in the co-management of their own local *blackfish* species. For these species, management tools applied at a local level may result in improved local fish stocks, and give direct benefits to the local community. Communities thus have the *incentive* to manage blackfish stocks, particularly where they have some form of 'use rights' to local spatial sub-units of the fishery. Such a sub-division of the fishery into management units would also provide the flexibility needed for effective local management.

These special characteristics of floodplain river fisheries suggest that some management activities will need to be undertaken at a catchment level while others occur at a local, community level. In between these two levels, there may also be certain situations where intermediate management units may be required. The more detailed sub-division of the fishery into management units is considered further in Section 4.

The question *who should manage?*, then, clearly does not have one single answer. This section attempts to show the process by which stakeholders at different spatial levels can determine who should take responsibility for different aspects of fisheries management. The factors influencing these decisions are discussed in the following sections, which focus on three key questions, as follows.

- Who has an interest or 'stake' in the fishery, and what is the nature of their interest?
- What activities or roles are needed to manage the fishery successfully?
- What skills and capacities do managers need to manage the fishery successfully?

Issues raised by these questions are then brought together in a table, which proposes a match of roles to three general categories of co-management stakeholders (community, intermediary groups and government) under two spatial management units (catchment and local). The match is based on who is most likely to have capacity at each level and illustrates that, although complex, it is possible to balance the hierarchical and spatial sharing of fishery management responsibility.

3.2 Who has an interest in management?

Many different groups will have an interest in the management of floodplain fisheries. The first step in understanding the 'who' aspects of management is to identify these different groups. The second step is to understand the nature of their interest, for example fishers interested in improving their daily food security have a different stake to landowners interested in improving the profitability of their water bodies. The power to make decisions in the fishery will vary among these different stakeholders. For example, power often lies with individuals who have a financial or political interest, but this may not always be in the best interests of the fishery - neither the resource nor the fishers. Understanding this helps predict how people will be affected by management changes and how they may respond to them: actions to avoid negative outcomes can then be incorporated into fishery management plans. The third step is to look at the relationships between the different groups: are they largely co-operative or competitive, or is any one group dependent on another?

This process is known as 'stakeholder analysis' and is a vital step to be taken

Management roles may be shared both hierarchically, between co-management partners and spatially, between different geographic sub-units of the fishery

Stakeholder analysis identifies who has a stake, the nature of their interest and clarifies the relationship between the groups

alongside investigating the more technical aspects of fishery management. Since conditions change regularly as the years go by, stakeholder analysis should be seen as an ongoing process, rather than a once-only activity at the beginning of a project or new management intervention.

An indicative list of stakeholders who might have an interest in floodplain fisheries is given in the table below, along with a description of their potential interests in the fishery. The list provides only a very general picture: any stakeholder analysis of an actual fishery would have more clearly defined categories with more detailed, specific information on the stakeholders groups, their interests and relationships. The activities of each of these groups will have an important effect on the success

Community

Care should be taken when using the word 'community'. This will very rarely comprise only a single group of fishers, all with the same interests. More often, floodplain communities will include some households involved in fishing and other households employed in alternative floodplain activities such as farming. Within the fishing households, some may be full time, while others may be part time. Further differences may exist in social, cultural and financial status between households. There may thus be widely different interests between the members of any single community.

Floodplain fishery stakeholders and their interests

Stakeholder group	Nature of interest in the fishery
Individual fishers as part of a wider fishing community	*Improved food security or income* Fishers relying on floodplain fisheries for food and income have a very close interest in how the fishery is managed. Their activities - how they fish, where and when - may be directly affected by management, and poor management will reduce the quality, quantity and value of their catch. *This group is very broad: a full stakeholder analysis would divide the fishers into smaller categories, for example full-time and part-time fishers.*
Resource managers within floodplain communities	*Improved resource management / continuation of community traditions* Floodplain communities often have traditional methods for managing natural resources. For example, a village committee with elders and experts may set and enforce rules defining the use of natural resources such as forestry, grazing land and fisheries. Even where these institutions may not traditionally deal with fisheries, they may be a recognised authority which could potentially take a role if a community's responsibility for its fisheries increased.
Landowners, lease-holders	*Improved income* People in this group have the power to determine access to the parts of the fishery which they control. As flood waters decline, remaining water bodies within recognised land boundaries may be controlled by the owner of that land. Similarly people who have control of the lease for a large water body can admit and exclude fishers on their terms. People in this category may or may not be actively involved in fishing, but they are usually a very powerful group in a fishery
Associated industries, e.g. fish processors and marketers, nursery owners, people who provide credit, or owners of large fishing gears.	*Improved income* Many individuals are involved in processing and marketing fish catches. Since their business relies on the continued supply of fish, they have a close interest in the management of the fishery. People in these industries may be family members of fishers, co-operatives, or 'middle men' who organise the sale of fish in distant markets. They may even be based in cities, rather than on the floodplain. *As for the fishers, a full stakeholder analysis would probably separate this broad grouping into smaller categories.*
Intermediary organisations, e.g. NGO's and projects	*Alleviation of poverty / research / improved resource management* Organisations that are neither government nor local communities operate in rural areas. They usually have a remit to reduce rural poverty and this role may often extend to issues in natural resource management, including fisheries and community development.

Stakeholder group	Nature of interest in the fishery
Administrative levels of the government's Department of Fisheries.	*Improved productivity / income generation / alleviation of poverty* The Department of Fisheries represents the government in the management of fisheries. They will develop and implement policies to cover international responsibilities and national aims for fisheries. All administrative levels of government have a role in floodplain fisheries. The levels may include: • local officers who interact directly with fishing communities, mostly involved in monitoring, enforcing government legislation etc.; • regional level officers who coordinate local levels and have a remit for larger geographical areas; and, • the Minister who has ultimate responsibility in the government for fisheries development, sets national policies in line with national objectives and international responsibilities etc.
Administrative levels of other government departments	*Improved productivity /income generation / alleviation of poverty* Other government departments, including agriculture, forestry, environment engineering and transport may each be responsible for natural resources on the floodplain in ways that overlap with the fisheries

of any management approach taken. Awareness, consultation, and ideally the full participation of all stakeholder groups will increase the likelihood of successful management.

3.3 What types of activities are needed for floodplain fishery management?

Successful management requires that stakeholders take responsibility for a range of roles. Eighteen roles have been identified in Figure 3.2, although the list is not exhaustive. It is not expected that any one stakeholder group can do all of the roles or that each of the roles will take place at each level of spatial

management unit. Fortunately, by using a co-management approach, the roles may be shared between many stakeholders, distributed according to who is most able to achieve them. Each of the roles is briefly discussed below.

Establish management objectives

As discussed in Section 1.2, fisheries may be managed for a wide range of objectives, and different stakeholders will often have different objectives. While government may wish to impose a general goal of sustainable resource use, the detailed specification of *local* objectives must be made by those local partners responsible for the

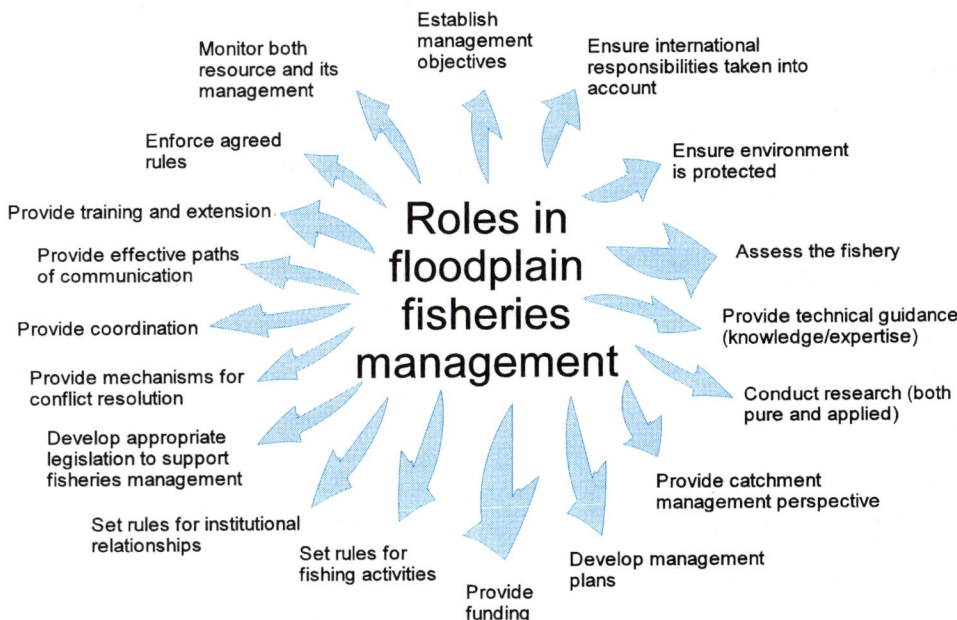

Roles in floodplain fisheries management

- Monitor both resource and its management
- Establish management objectives
- Ensure international responsibilities taken into account
- Enforce agreed rules
- Ensure environment is protected
- Provide training and extension
- Provide effective paths of communication
- Assess the fishery
- Provide coordination
- Provide technical guidance (knowledge/expertise)
- Provide mechanisms for conflict resolution
- Conduct research (both pure and applied)
- Develop appropriate legislation to support fisheries management
- Provide catchment management perspective
- Set rules for institutional relationships
- Set rules for fishing activities
- Provide funding
- Develop management plans

Figure 3.2 Management roles required for effective management of floodplain fisheries

*No one
stakeholder
group will be
able to take on
all of the roles*

management unit, within the principle of sustainability. Local people will not contribute effectively to the management of the fishery if they have not taken part in establishing objectives. The objectives at each level should be complementary. Where differences do exist, they cannot be ignored - stakeholders at the different levels should discuss the conflict in objectives and reach a compromise.

Ensure international responsibilities are taken into account

As floodplains are part of larger river systems that may cross country borders, management of their fisheries needs organisations capable of making decisions on wide geographical and sometimes political levels. Certain specific management tools, such as species introductions, may be constrained by international agreements.

Ensure the environment is protected

A healthy environment provides the basis for the high productivity of the fishery, but is highly vulnerable to overuse and degradation. Many floodplain activities have the potential to alter hydrology (i.e. the quality, quantity, timing and duration of annual floods) and water quality (e.g. pollution from agricultural pesticides or industrial effluents). Countries are often bound to protect natural resources through international agreements.

Assess the fishery

Management must be based on an understanding of the floodplain fishery, i.e. the environment, the fish, the fishing practices and the stakeholders. Assessments of flood patterns and migratory whitefish must be made at a catchment wide level while individual fishing grounds and black fish must be assessed at the local level. Tools such as stock assessment models may assist the technical appraisal of a fishery, while a range of rural appraisal methodologies (e.g. participatory rural appraisal, stakeholder analysis) may provide information on stakeholder involvement. Assessment of the fishery can also be undertaken by members of the fishing community on the basis of their own fishing experience.

Provide technical guidance (knowledge / expertise)

Floodplain fisheries are complex: many different fish are caught by many different gears, used by many different people. Technical understanding of this complexity may be gained through both traditional knowledge (often detailed and specific to a particular area) and scientific knowledge (important for a catchment perspective). Technical guidance contributes to the assessment of a fishery and the development and implementation of a management plan.

Conduct research - pure and applied

Research may contribute to the broad understanding of the floodplain system (e.g. pure scientific research of floodplain ecology or hydrology) or may be part of the daily management of a fishery (e.g. adaptive management where managers 'learn by doing' and so increase their understanding and ability to manage). Floodplain fisheries are often well understood from a technical point of view, but poorly understood with regard to social and institutional issues which also determine the success of management.

Provide a catchment perspective for management

Since the quality, quantity and timing of flood water provide the basis to floodplain fisheries production, managers must consider floodplains as part of entire river systems. Large-scale interventions such as dams, flood control measures and the cumulative effects of many small scale interventions carried out at local level may all affect floodplain fisheries. Catchment managers must balance advantages for one sector against the potential impacts on another (most often the fishery). Clearly, this is a cross-sectoral activity, so co-ordination and communication are critical for success. Migratory whitefish stocks also cross many community fishing grounds and thus require management at a catchment level.

Develop management plans

A management plan for a fishery may specify the objectives of management, the tools by which these objectives may be achieved, and the responsibilities of the different partners in the management process. The full development of a management plan may require each of the following steps:

- identification of management units (see Section 4.1);
- stakeholder analysis (see Section 3.2);
- selection of management objectives (see Section 1.2);
- selection of management tools (see Section 4.3);
- assessment of stakeholder capacity (see Section 3.4);
- collective agreement on responsibilities of each stakeholder (for example, see Section 3.5); and,
- development of a legal and policy framework for management (see Section 3.5.1).

Setting rules for fishing (i.e. who can fish, which species, where, when and how)

The technical basis for fisheries management is the set of rules defining who can fish, which fish they can catch, and where, when and how they can catch them. The high variability of floodplains (water, fish, fishing gears, fishers) means that there are few rules which are universally applicable for all parts of the fishery. A flexible approach to selecting management rules is therefore essential. To improve the likelihood that fishing rules will be obeyed, they should be locally appropriate and made by the people who will be governed by them. Decisions on 'who can fish' are very important in terms of wider management objectives for the distribution of benefits. Appropriate setting of access rules provides a powerful way to direct benefits to a targeted group and ensure that vulnerable groups are not excluded.

Set rules for institutional support of fisheries management

An important, and often overlooked, part of fisheries management is the analysis of stakeholders, their inter-relationships and their potential influence on outcomes of management. As floodplain fisheries management must consider national, catchment, and more local elements of the resource, it requires the involvement of stakeholders at all of these levels. It should be clear and generally agreed which stakeholders will have responsibility for which roles. When different groups need to work closely together, it is helpful if the nature of their reationship is clarified.

Develop appropriate legislation to support fisheries management

Formal legislation should be used to give authority to the co-management partners for the management of their fishery. Legislation may provide critical recognition and support, particularly when attempting to limit access to the fishery. However, since the formal law-making process is slow and unwieldy, it will never be flexible enough for the year-to-year management of each local fishery. Formal laws on mesh sizes or small portable gears may also be almost impossible to enforce from above in dispersed, rural fisheries. National legislation for floodplain fisheries should therefore aim to provide an *enabling framework* within which more detailed, locally appropriate management can take place rapidly and independently, but still with the full backing of the law.

Provide mechanisms for conflict resolution

Fishery managers will often need to resolve conflicts, either between different fishers, or between the fishery and the other sectors that have a claim on floodplain resources (e.g. agriculture, transport, aquaculture etc). Conflict resolution involves three steps: discussion, adjudication and enforcement. These steps can take place formally, for example in a court with a judge deciding some legal penalty or informally, for example in a village meeting chaired by an experienced and respected fisher who decides on some social sanction.

Co-management means that the roles can be shared between stakeholders

Co-ordinate

Floodplain fishery management involves people and decisions at many different administrative levels (national, catchment and local), from many different sectors and from many floodplain communities. Effective co-ordination of all of these stakeholders will be a vital role to ensure that activities and responsibilities are complementary and do not conflict with each other. Experience in co-ordination may be quite limited in fisheries and so it is important to establish an agreed system within and between relevant stakeholder groups.

Communicate

Effective communication will build trust between stakeholders and encourage their continued participation in the co-management partnership. Exchange of information between stakeholders in floodplain fisheries is important to develop, maintain and improve fisheries management. Good pathways of communication are necessary both within and between organisations. Many different methods of communication may be used, for example, posters, regular meetings, workshops, newsletters, study tours etc.

Provide training and extension

To be successful, the people involved in fisheries management will need operational, technical, social, financial, economic and management skills. Usually, the co-management team will not have all of these skills, so training will be required. Training can either be formal or informal, and may include focused workshops, visits, conferences, individual courses or on-the-job experience.

Monitor

Monitoring is an essential role, needed to assess both the state of the fishery and the effectiveness of management. Fish stocks, fishing activities and outside environmental influences should thus be monitored (see Section 4.4), in addition to the performance of the various stakeholders in carrying out their management roles. Feedback should be given to the stakeholders at regular intervals both to maintain their commitment to the co-management process, and to improve their effectiveness in their roles.

Enforce

Rules are made to govern fishing activities so that fisheries management objectives are met. To be effective, rules must be enforced and a system must be established to deal with rule-breakers. The system may either be based in the legal system with fines being the main form of penalty, or be community based with a range of penalties from short term exclusion from a fishery, through to complete social exclusion. It is often beneficial to have penalties of variable severity, so that first offenders may be penalised less heavily than the more regular lawbreakers.

Fund fisheries management

Fishery management will require funding for a wide range of different activities, such as training, producing posters and newsletters, collecting monitoring data, resolving disputes, developing capacity and so on. Some management tools such as stocking or habitat restoration will also have capital or labour costs. Over time such costs should increasingly be recovered from the fishery itself, usually by charging fishers in some way for their access to fishing. This 'cost recovery' will be most successful where the access rules for the fishery are widely understood and agreed, and a transparent financial system is established to prove that funds are being used in the agreed manner. The use of credit schemes as a method of supporting fisheries management needs to be investigated. Credit may be particularly relevant where communities are taking on new roles and need to develop different skills.

3.4 What stakeholder capacity is needed for management?

To decide who should take responsibility for these many different roles, it is necessary to consider who *could* take responsibility, and also who would have the *incentive* to take responsibility. This section thus deals with the 'capacity' that stakeholders need to successfully take on the various management roles in the fishery. In the following sub-sections, four different types of capacity are discussed - resources, skills, rights and motivation - and questions are asked to help clarify whether the necessary capacity is available.

3.4.1 Resources - people and money

Fishery management requires both individuals to carry out the different roles and money for them to work effectively. Some management tools (e.g. stocking) require capital investment.

> ? How many people are needed to carry out each role? Do stakeholder organisations have enough staff or members?
> ? What are the costs associated with each role (e.g. for staff time, travel, training, equipment, monitoring etc.)? Is the budget adequate?

3.4.2 Suitably trained people

Fisheries management needs people with skills in many areas, from technical skills for the assessment of the fishery, through to social and management skills for encouraging co-operation and reducing conflicts. Co-ordination and communication skills are also very important in managing such complex resources.

> ? What kind of skills are needed and does the organisation have people with these skills?
> ? How do people within an organisation communicate with each other?
> ? How will information be passed between people in different organisations?
> ? What kind of training would be most suitable, and cost effective?
> ? Who could provide the training?
> ? How long would it take, who should be trained, and who would pay?

3.4.3 Rights - recognised roles, responsibilities and the right to manage

With so many stakeholders involved in floodplain fisheries, it is important that everyone understands and agrees who is to perform each role. The rights and responsibilities that go with each role must be clear at all levels, from villages up to government departments, and include any local NGOs or projects. Conflict in fisheries is often caused by confusion about who does what. Without a recognised right to manage, the actions taken by a group of stakeholders can be ignored or challenged. Recognition can either be formally written, such as under a legal agreement (e.g. a lease), or exist informally in the case of traditional community management rights.

> ? What roles have been assigned to different stakeholders and by who?
> ? How has that responsibility been recognised and recorded (e.g. national legislation, letter of agreement or informal conclusions of meeting)?
> ? What will happen if conflict arises between stakeholders over who has responsibility?

3.4.4 Motivation

Without appropriate incentives, people will not want to be actively involved in fisheries management. In government departments, incentives may be salary, promotions, training etc. For fishers, incentives may include early benefits from improved management of the fishery, in addition to secured access to the fishery, long term control of the resource, opportunities to learn new skills, tangible benefits such as more fish, larger fish or social recognition.

> ? What are the incentives for different stakeholders, and are they adequate?
> ? What are the disincentives for involvement, and can they be overcome?

For each co-management partnership, then, the decision on 'who should manage' should take account of the

The capacity of stakeholders will help determine who should take responsibility for each role

Capacity can be thought of under four headings: resources, skills, rights and motivation

capacity of each stakeholder group (Section 3.2) to fulfil each of the different roles (Section 3.3) successfully. If skills, experience, or resources are lacking, then such capacity must be developed. This may involve the exchange of skills between people, villages or government departments, or formal training through workshops or courses. External support from local development projects may also be vital in helping a community develop the necessary range of skills.

3.5 Matching stakeholders to roles in fisheries management

As noted earlier, no single group of stakeholders will have the capacity to take on all of the roles necessary to manage the fishery. The full combination of capacities may only be available in co-management partnerships involving representatives of different stakeholder groups at the appropriate levels. Which stakeholder should take on which role in each co-management partnership will depend on their respective capacities and other local factors. The following sections give general advice on the contribution that different stakeholders may make to each role. For this illustration, the stakeholders are grouped as government departments, local floodplain communities and other intermediary organisations such as development projects and NGOs. The section concludes with a potential match between these broad groups and roles at two levels of management unit (local and catchment).

3.5.1 Government departments

Although current trends in resource management are for devolution of responsibility to communities, the characteristics of floodplain fisheries mean that *governments will always have an important role*.

The administrative levels of appropriate ministries provides an *established nested structure* (e.g. national, regional, district....etc.), which could support the layers of management units in Figure 4.1. Governments are in a good position to *coordinate activities* across different sectors. They can also identify areas where research could improve management and *apply for national and international resources* through projects.

Due to their overview of fisheries in a country, governments are also in a good position to initiate the *identification of potential management areas* (catchment, and various local units) for floodplain fisheries. Governments can also set up the framework and *arrangements for giving authority* to these various management units. This is an important part of the strategy for floodplain fisheries as outlined in this document.

Fisheries Departments are responsible for making *national policy* and this will influence the legal system governing fisheries management. Fisheries law plays an important role in the management of fisheries in each of the units illustrated in Figure 4.1 from national, through catchment to local units. Only national law is capable of devolving responsibility to lower management levels. An important role for government is thus to *provide a legal framework* that enables appropriate institutions in each of the management units to take responsibility for their section of the fishery.

An important consequence of governments' giving authority to various management units is their continued involvement in *conflict resolution*. Challenges to the established system must be met and government has the capacity to support valid managers against the claims of others wishing to take control of the resource. Government must ensure that there is a framework for conflict resolution within fisheries, and between fisheries and other sectors. This must be supported by appropriate policies, institutions, laws and information flows. Although the actual resolution of fisheries conflicts at village level will usually best be done by suitable supported village institutions, national legislation may be required to empower them to do this.

Clearly government involvement is necessary where catchments cross national boundaries. In addition, governments often have *international responsibilities* for the management of natural resources and the environment which will have implications for fisheries management.

Staff of fisheries departments often have considerable *technical knowledge* and may be actively involved in research. Therefore, these departments can *provide advice* and take responsibility for *sharing information* that will improve the management of the fishery.

3.5.2 Floodplain communities

Floodplain resources comprise many different fishing grounds of various sizes, including river channels, permanent water bodies, and seasonally flooded pools and shallows. Their hydrology and vegetative cover differ and vary with the season. This determines the types of gears which may be used and the vulnerability of both blackfish and whitefish. Decisions on how to manage these individual fishing grounds are beyond the capacity of any national organisation. They are more appropriately made by local communities. Although the legal and policy framework must be agreed with a national organisation such as a Fisheries Department, most *decisions and activities* should be made at a very *local level*.

However, for a range of reasons, some communities may not be able to make such decisions effectively. A list of conditions which may improve the likelihood of successful community participation in management is given in Box 1. Situations where these conditions are not met and cannot be developed or encouraged will present a real constraint to the local management of the fishery. The successful involvement of communities depends particularly on the unity of the stakeholders and the strength of their systems of authority. A divided community with a weak authority system is less likely to successfully manage their unit of the floodplain, than a *unified community* with *clear, strong leadership*.

Community involvement in fisheries management also requires a *great deal of trust among stakeholders*. This requires time to develop, particularly where previous relationships have included some degree of conflict. Close involvement of a development project or another intermediary group will often be required to provide the incentives for community stakeholders to become actively involved. Without this stimulus, people may be reluctant to participate, particularly if their previous involvement has been limited and they cannot see how they will benefit.

3.5.3 Intermediary organisations

This category covers a range of organisations, such as non-governmental organisations (NGOs), international projects, aid agencies, extension and development projects. Such organisations are often active in rural areas and may help to improve linkages between government and communities.

These groups often have a *poverty focus*, which includes resource management and may extend to environmental protection. Projects are often *research-based*, aiming to understand the nature of the resource to *assist management*.

People in these organisations often have *specific skills in training, extension, and communication* which can assist both government and local communities with their responsibilities for fisheries management.

In some circumstances projects or agencies may provide *initial support*, such as developing skills or providing credit, which can help communities and governments *build their capacity* to manage. Projects may be critical to developing the meaningful involvement of communities in management by helping to clarify roles, introducing management methods and procedures, encouraging stakeholders to take on new management responsibilities, helping to identify the benefits of participation, and reinforcing relationships between stakeholder groups.

Intermediary organisations may help governments and floodplain communities to develop their management capacity

Devolution of
responsibility to
floodplain
communities is
not a simple
task: a range of
resource and
community
characteristics
provide guidance
to where the
chances of
success will be
higher.

Box 1: Conditions that improve the chance of successful community involvement in fisheries management

- The **boundaries** of the management unit must be clear and of a manageable size. This includes both the *physical boundary* of the fishing ground, e.g. an agreed area of a floodplain such as a dry season water body, and *who* is part of the management unit, e.g. a list of legitimate fishers and the membership of any management committee and its structure.

- The 'community' should live near the fishery and have a **common approach** to shared problems. A community with previous experience of solving problems facing its members and that has a shared understanding of key objectives, will have more chance of successfully meeting the challenges of managing a fishery than a community with many internal conflicts.

- It is helpful if the community has an **existing organisation** with management responsibility, even if not specifically covering fisheries, as this provides a structure for the introduction of any new management interventions. Previous experience of managing natural resources within a community provides a good basis for stakeholders to take on responsibility for fisheries management.

- Individuals are more likely to participate when it is clear to them that the **benefits exceed the cost** of their involvement. Therefore, it is very helpful if benefits are demonstrated early in the process of developing stakeholder capacity - this is a great incentive for future involvement.

- Rules that specify who can fish, how, where and when must **reflect local conditions** (i.e. they must be appropriate for a multi-species, multi-gear fishery on a seasonal floodplain exploited by many different types of fisher).

- Rules are **best made by, or with the cooperation of, the individuals** affected by them. This includes detailed fishing rules as well as rules governing who can make and change the arrangements guiding the management of the fishery.

- Managers need to **monitor** both the state of the fishery and the activities of fishers, including any rule-breaking. Monitoring should either be carried out by the fishers themselves or at least people accountable to them.

- Communities should set up a **system of penalties** to deal with people who break rules. The system should include a mixture of light and severe penalties to deal with different levels of rule breaking and individual circumstances. Penalties should be seen to be given out. This gives confidence to stakeholders that monitoring is effective and that rule-breakers are penalised. These are both key factors in encouraging rule compliance.

- Communities should establish ways of **resolving conflict**. Mechanisms should be fast and low cost, relying on both formal (e.g. law courts) and informal (e.g. committee meeting) methods.

- Communities must have **external recognition** of their right to manage (e.g. government legislation giving tenure or government policy delegating responsibility to a well defined management unit).

- Management should be supported by a **nested arrangement of organisations** such as the system of river fisheries management units in Figure 4.1. All types of local floodplain management units are grouped within their river catchment, and all catchments will be grouped under a national organisation.

- There should be a core group within the community that takes **leadership responsibility** for the management process. Individual fishers may need incentives to commit time, money and effort into fisheries management.

- **Communication between government and the community** requires a joint body to be established. Membership should include representatives from both stakeholder groups (community and government) and should have a remit to monitor progress, resolve conflict and reinforce local decisions.

Adapted from:
Ostrom, E. (1990). *Governing the Commons. The Evolution of Institutions for Collective Action.* Cambridge University Press, Cambridge, 280 pp.
Pinkerton, E. (Ed) (1989). *Cooperative management of local fisheries. New directions for Improved Management and community Development.* University of British Columbia Press. Vancouver, 299pp
Pomeroy and Williams (1994). Fisheries Co-management and small-scale fisheries: a policy brief. International Centre for Living Aquatic resources Management, Manila 15pp.

3.5.4 Illustration of who might do what

In summary, the table below illustrates how communities, intermediary organisations and government departments could contribute most usefully to the various management roles at both local and catchment levels.

It is clear from this table that, in local management areas, floodplain communities can take a very active role with governments providing the supportive framework. At the catchment level, governments will more often be the lead organisation. At both levels of management unit, the intermediary organisations may provide support to both governments and communities.

It should be clear by now that *devolution of fisheries management to local co-management partnerships is not a simple option*. It requires legal, technical, financial, social and administrative support. Ironically, it also requires a strong central government, committed to the principle of decentralisation. The proposed hierarchical and spatial sub-division of responsibilities is designed to provide a structure, based on the characteristics of the resource, that will allow decentralised local partnerships to gradually take on full responsibility for managing the fishery. The following section describes in more detail, *how* resources may be sustainably managed.

Potential roles for different stakeholder groups in local and catchment management units

	Floodplain communities	Independent Organisations	Government Departments
Local Management Areas (VMAs and IMAs)	set objectives protect environment resource assessment technical guidance research decide management plans set fishing rules set institution rules conflict resolution co-ordination communication training monitoring enforcement funding	protect environment resource assessment technical guidance research decide management plans set fishing rules set institution rules conflict resolution co-ordination communication training / extension monitoring funding	international responsibility protect environment technical guidance research catchment management set institution rules develop legislation conflict resolution co-ordination communication training / extension funding
Catchment Management Areas (CMAs)	protect environment set institution rules conflict resolution communication monitoring enforcement	protect environment resource assessment technical guidance research catchment management decide management plans set institution rules conflict resolution co-ordination communication training / extension monitoring funding	set objectives international responsibility protect environment resource assessment technical guidance research catchment management decide management plans set fishing rules set institution rules develop legislation conflict resolution co-ordination communication training / extension monitoring enforcement funding

Co-management will present a major challenge both to government and other stakeholders: managers should start in a few local areas and build gradually on that experience

Management units should be selected to achieve the maximum overlap between the range of authority of the management group and the distribution range of a fish stock

4 How to Manage?

This section shows *how* floodplain river fisheries may be managed by their stakeholders. Section 4.1 describes how the resource may be sub-divided into different types of management units. Section 4.2 then provides guidelines on the strategic assessment of those units, showing the types of information needed to make useful management decisions. Menus of alternative management tools are then described in Section 4.3 for each of the different management unit. Finally, in Section 4.4., methods are given for monitoring floodplain fisheries to ensure the effectiveness of new management strategies.

Though this section focuses mainly on *technical* tools and assessments, it is re-emphasised that effective management will require far more than simply making a choice between say a reserve and a mesh size regulation. As indicated in Section 3.3, management requires a whole range of different roles. Only a few of these, such as research and fishery assessment contribute directly to the process of selecting technical tools. The majority of roles - including legislation, co-ordination, communication, enforcement and monitoring - are more intended to ensure that the rules are widely understood and supported by the different stakeholders, and also that they actually achieve the selected objectives for the fishery. Without this much wider management support, technical management will be worthless.

The technical proposals in this section should also only be read as *guidelines* for the management *process*, rather than as blueprint solutions for universal application. As emphasised in parts of Sections 4 and 5, the successful management of floodplain fisheries will need to develop gradually over a series of steps. Many important lessons will, no doubt, be learnt as new co-management strategies are applied. Managers should always make the best use of these lessons, rather than sticking rigidly to any pre-conceived format or plan.

4.1 Floodplain fishery management units

Section 3.1.2 introduced the idea of dividing the floodplain into a nested arrangement of fishery management units, and promoting the development of co-management partnerships in those units with the best opportunities for this type of management. The sub-division of a river system into management units should be based on the spatial interactions between the floodplain river environment, the fishing communities and the fish stocks. In practical terms, three main categories of management units may be identified: Catchment Management Areas (CMAs), Village Management Areas (VMAs) and Intermediate Management Areas (IMAs).

Figure 4.1 illustrates how a floodplain may be split up into these different units. All rivers should be expected to have many different potential VMAs, and maybe a few different IMAs. As discussed later, VMAs may sometimes be nested within IMAs. As illustrated on the left of Figure 4.1, a small river may be manageable with only a single CMA, while larger rivers may be better sub-divided into several CMAs. Rivers straddling international borders may even have some CMAs split between two countries (increasing the difficulties of successful management in those parts of the catchment).

Each of these management units will require different management contributions from each of their stakeholders, as discussed in Section 3. They will also require the use of different management tools, as discussed in the following Section 4.3. The following table briefly summarises the justification and requirements for the different management levels. Guidelines for identifying the different units are given in the following sub-sections.

4.1.1 Identifying Village Management Areas (VMAs)

Fishery management goals are most likely to be achieved when management rules are well adapted to both the physical characteristics of local resources and to the social priorities of local communities.

Floodplain fishery management units

Management Unit	Justification and management needs
National / International	Large river systems may flow across many different parts of a country, or even between two or more countries. Their management requires institutions capable of making decisions and resolving conflicts on wide geographical, and sometimes political, levels.
Catchment Management Area (CMAs)	Floodplains are only a part of the overall river system. Many different sectors may compete or the valuable resources of the floodplain. Changes to the quality, quantity and timing of the flood due to upstream activities can all cause negative impacts on floodplain fisheries. Maintaining the joint productivity of fisheries and other sectors thus requires their co-ordination at a catchment-wide level. Whitefish stocks also migrate around the whole catchment, and require management at this scale.
Village Management Areas (VMAs)	The high local variability of floodplain river systems and their fisheries means that no single approach to managing each of the smaller units in a floodplain will succeed. Dividing the floodplain into a number of small, local units gives the flexibility needed for effective local management. Allocating use rights to these small management units may also give local communities the incentive to manage their local blackfish stocks for sustainable long-term benefits.
Intermediate Management Areas (IMAs)	In some parts of the floodplain, the geographical distribution of villages and waterbodies may mean that the catches in each village are more than usually dependent on the activities in their neighbouring villages. In such situations, co-operation between the different villages may be required to achieve management goals. This requires particular skills in communication and co-ordination, therefore IMA management may be expected to be more difficult than VMA management.

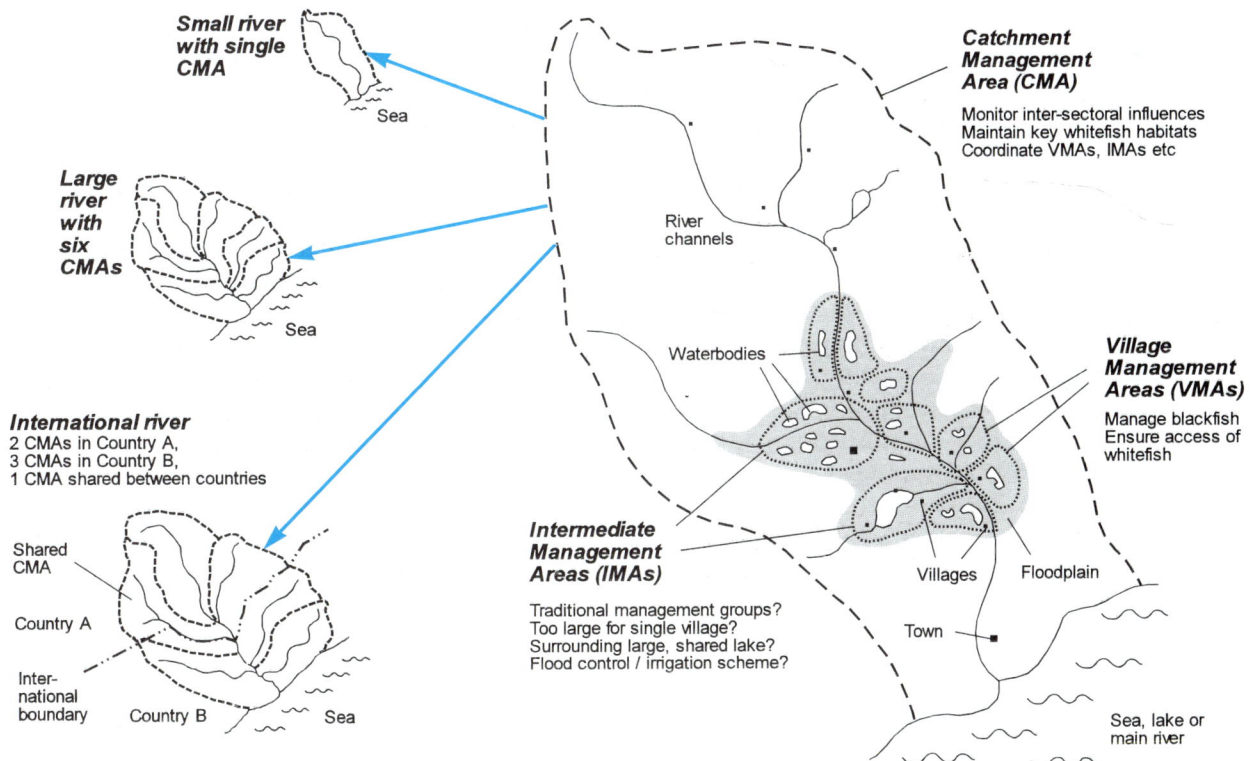

Figure 4.1 River fishery management units

VMAs provide
the strongest
management
opportunities
wherever fishing
communities
have traditional
control over local
waterbodies
within areas
small enough to
manage
effectively

The selection and enforcement of management rules may thus be best achieved by the local community, taking advantage of their intimate knowledge of their resources and their capacity for mutual monitoring and enforcement. Local village management areas (VMAs) will therefore provide the best starting point for most floodplain fisheries management strategies.

VMAs should be selected to achieve the maximum overlap between the range of authority of a social group (e.g. a village), and the distribution range of a blackfish stocks. Managers thus need information on the spatial distribution of four items: waterbodies, fish, fishing and existing management 'institutions'. Regional data on some of these subjects may be available from existing records of the fisheries departments and planning agencies; local data will need to be collected by interviewing key members of each fishing community. Training on effective community research techniques may be required for this process. The distribution and behaviour of fish species will usually be the most difficult information to determine, and it may be necessary to assume that floodplain regions will have some local blackfish stocks wherever there are significant dry-season waterbodies.

VMAs may be identified by holding discussions in villages, based on the checklist of questions in Section 4.2, and bearing in mind the 'Who should manage?' issues raised in Section 3. Localities suitable for use as VMAs may also already be known by local fisheries officers. The best prospects exist in those fishing communities which have had traditional control over their own waterbodies, especially when their control is recognised and endorsed by government. There may also be some advantages in adopting local government boundaries, to take advantage of existing administrative abilities and systems of authority.

4.1.2 Identifying Catchment Management Areas (CMAs)

One or more 'Catchment Management Areas' (CMAs) will also be required for all river systems. CMA-level

management would have three broad purposes:

- monitoring and management of the impacts on the fishery from other sectors;
- co-ordination of management activities in local VMA and IMA units, and communication of the successes and failures of alternative approaches between local units; and
- management of migratory whitefish stocks.

Where possible, CMAs should be identified on hydrological grounds, as the geographic area from which water (and pollution etc.) drains into the river system. Where catchment boundaries do not overlap with administrative boundaries, some boundary adjustments may be needed to enable the participation of existing management agencies.

Small rivers may be manageable with only a single CMA wherever the managers can coordinate activities across the whole catchment area. Larger rivers may be more effectively divided up into two or more CMAs, either for hydrological reasons (e.g. where the river system has two or more discrete floodplain regions) or to divide the river catchment between separate administrative authorities. Since whitefish (and water, and pollution etc.) may still move between CMAs, multi-CMA rivers will require co-ordination by a river management forum. In the very largest rivers, such as the Ganges or the Mekong, international fora (such as the Mekong Secretariat) will be necessary.

4.1.3 Identifying Intermediate Management Areas (IMAs)

Management opportunities for blackfish will be greatest in 'bottom-up' VMA-level management units. Whitefish may be primarily managed in the more 'top-down' CMA-level units. Between CMAs and VMAs, however, there may also be a range of 'Intermediate Management Areas' (IMAs), whose management needs and opportunities depend on the spatial relationships between waterbodies and communities. IMAs would include single large waterbodies (floodplain lakes) fished by two or more surrounding

Simpler
management
tools are
required for
IMAs, due to
the increased
difficulties of
roles such as
monitoring,
communication,
coordination and
enforcement in
these larger
areas

Problem	Management Solution
A waterbody (e.g. large lake) is shared between a number of villages, giving significant overlap of *blackfish* stocks. Independent management by one village may be negatively affected by the actions of the other villages.	No spatial VMAs created, but village leaders control fishing activities of their own local people. IMA created with representatives of each village, to negotiate and agree rules to be followed at IMA level (i.e. throughout the whole waterbody).
A floodplain comprises multiple waterbodies, not closely associated with individual villages. Waterbodies are too remote for village-based monitoring and enforcement.	No VMAs created. Waterbodies are managed at IMA level using less ambitious management tools, and with representation from surrounding villages or districts with interests in the fishery.
Local hydrological modifications such as flood control schemes or impoundments disrupt *whitefish* migrations, to the cost of all of the impounded villages, or create fishing advantages for some villages and disadvantages for others.	VMA managers control fishing activities within their own local waters. An IMA is created with representatives of each village, to resolve local disputes and negotiate with other sectors.

villages; and single villages or towns lying alongside a large, multi-waterbody floodplain system. Depending on the fishery problems at hand, IMA units may either provide an 'umbrella' over two or more VMAs, or completely replace them at the base of the fisheries management hierarchy. The following table illustrates the types of problems where IMAs may prove useful.

Simpler management tools are required for such IMAs, due to the increased difficulties of roles such as monitoring, communication, co-ordination and enforcement in larger areas. As for VMAs, management of these units will usually be easiest wherever traditional management institutions already exist. Where they are currently absent, new management institutions should only be proposed when substantial levels of management support can be allocated. In the first place, managers should concentrate on establishing VMA and CMA-level management, and work on IMAs only where some institutional structures already exist.

4.2 Strategic assessment of the management units

4.2.1 Information requirements

Four categories of information are required to understand a fishery management unit and guide the local selection of technical management tools: environmental conditions, the distribution of fish stocks and fishing practices, and the 'institutional arrangements' currently in place for managing the fishery (see checklist in Box 2).

At the VMA and IMA level, much of this information may be collected directly from the local fishers collaborating in the co-management partnership. At the CMA level, managers will need to rely on more regional data sources, and on combining the knowledge of local units in different parts of the catchment. Though fishing communities may have only limited knowledge about wider fish migration patterns (e.g. about the full range of whitefish movements), they will usually know about the timing and direction of fish movements within their own fishing grounds and about their dry season survival locations. Most importantly, they will also usually know what actions would sustain their local catches, though they may never before have had the incentives to follow these actions. From a management perspective, the answers to the following critical questions should be found by the successful application of the checklist:

- Are fish stocks relatively stable or in decline (smaller sizes, harder to catch)?

The appropriate management strategy for each fishery unit depends on the objectives selected for it and its current performance against those objectives

Box 2. Checklist of data required for effective management of floodplain river fisheries

ENVIRONMENT

- Regional and local **maps of waterbodies**, including:
  Waterbody **names** & **types** (rivers, lakes, floodplains); water flows
  Minimum dry season **water depths** (where *could* fish survive the dry season)
  Villages and their associated **fishing areas**
- Regional **maps or satellite images indicating sectoral resource use** (potential catchment influences on fisheries)

FISH

- Main **fish species** caught in each waterbody type?
- Location of **spawning areas** of main species?
- **Accessibility** of village fishing grounds to fish from main rivers (natural / fishing barriers)?
- Which waterbodies do fish **survive** the dry season in?
  Are local fish stocks fished out in the dry season? If not, why not?

FISHING

- Matrices of **relative catch values** between the main fishing seasons and gear types
- For the gear/season combinations (cells in the matrix) producing the largest catches:
  Which **species** caught?
  What **variability in timing**? (Identify concentrations of profit for particular gears)
  Access restrictions influencing allocation of surplus?
  Group or individual fishermen? **Immigrant** fishermen? Other categories?
  Details of **gear operation** (team composition, gear costs, share distribution / payment)?

INSTITUTIONAL ARRANGEMENTS AND OBJECTIVES

- What **regulations** are there? **Who** makes each one, and **why**?
  Species? Seasons? Gears? Access? Places? Allocation of fishing spots?
- **Relationships** between the rule making bodies? (**Formal** and **informal**...?)
- **Monitoring** of regulations
  Who does it? How is it done? How effective is the monitoring?
- **Enforcement** of regulations
  Who does it? How is it done? What are the penalties for breaking?
  How often are penalties applied?
- **How long** have the regulations been established? Any changes?
- Fishing **conflicts** or other problems?

- Which stocks are declining and are they blackfish or whitefish?
- Which permanent local waterbodies do *blackfish* survive the dry season in (how could they be protected)?
- Can *whitefish* access local fishing grounds from the main rivers (could management improve accessibility)?
- Could the distribution of catch between different stakeholder groups be improved?
- What changes in fishing (or water use) practices could help stop the decline in stocks or improve fisheries outcomes in some way?

The appropriate technical management strategy for each fishery unit then depends on the capacity of the co-management partners, the objectives selected for the unit, and its current performance against those objectives. From a quick examination, the most heavily exploited fisheries may be distinguished by the following characteristics:

- Fishing gears with small mesh sizes (e.g. less than 4-5cm stretched mesh);
- Small fish in the catch (e.g. less than 10-15cm in length, including both small fish species and small specimens of large species); and,
- Many competing (chasing) fishing gears, used during the (inefficient) high water period.

Where measures appear necessary to improve the outcomes from the fishery, managers should select from the lists of alternative management tools proposed in Section 4.3. The selection of tools should take into account their likely impacts on the fish stocks and the fishing gears currently used in the fishery (and their owners), and on their acceptability to local stakeholders. Before moving on to the selection of technical tools, the following two sub-sections describe the types of impacts which should be expected for fish stocks and fishers.

4.2.2 Predicted impact of management tools on the recovery of fish stocks

The lists of management tools given in Section 4.3 indicate their *theoretical benefits* for fish stocks in each type of management unit. Such benefits are both based on common sense, and supported by the technical experience reported in Part 2 of this paper. Where blackfish stocks are locally depleted, for example, reserves may provide a good means of restoring their numbers. Where whitefish stocks are seen to decline following the introduction of barrier traps, the removal of the barriers may restore those fish species. The actual impact of the reserves or barrier regulations on the recovery of the preferred species may, however, be virtually impossible to predict in advance, due to the complexities of local conditions. For this reason, these management tools are recommended as simple approaches, which should always be followed up with *monitoring* of their impacts and further *adaptive management* as and when required (see Section 4.4).

4.2.3 Predicted impact of management tools on the catches of different fishing gears and their owners

Though the impact of management tools on the recovery of fish stocks may be difficult to anticipate, their impact on the *distribution* of catches between gears will usually be much more obvious. As described in Section 2.4,

due to the competitive interactions between fishing gears, technical management approaches such as gear closures, closed seasons, and mesh/fish size limits always have benefits for some gears, and losses for others.

A basic understanding of the likely impacts of alternative management tools can be gained from simple information on the relative catches of each main *fish species* taken by each main *gear type* in each *season*. For VMAs, this information should be available from within the fishing community involved in the management decisions. For higher level management units, such information may be obtained either from catch sampling programmes, or from interviews with fishers.

With an overall knowledge of the catch pattern in the fishery, the expected gains from the management strategy for some gears can be easily balanced against the likely losses for others. This idea is illustrated in Figure 4.2 for a simple example with only two species, two gears and two seasons. Each of the three management strategies would have benefits for the users of gear A, and corresponding losses for those using gear B. While any of the three strategies may be useful for this fishery (especially if the catch of species A is the main objective), they clearly should *not* be implemented without serious consideration of their impacts on the users of gear B.

A real floodplain fishery would of course usually have more than two gears, along with many fish species and more than two seasons. The types of interactions between the gears would, however, remain the same, and could at least be identified as above for the main gear types and species. This simple, but multi-species, multi-gear approach should always be considered during the development of new management plans for floodplain river fisheries. The adoption of a given strategy should then be discussed between all the stakeholders likely to be affected (particularly those fishers relying on 'gear B').

4.3 Fishery unit management plans - selecting from a menu of technical tools

This section describes the alternative technical tools which may be used by managers to achieve sustainable benefits from the fishery. As mentioned earlier, such technical tools are only one component of a *fishery management plan*. The plan should also list the management objectives for the fishery and clarify each of the roles needed for successful management, such as communication, monitoring, enforcement and so on. It should also identify which stakeholders have responsibility for each role (Section 3.5), and be revised as their understanding, experience and capacity increases.

No single technical management tool will answer all the objectives of management. A combination of tools should thus be selected for each fishery management unit, as appropriate to the local situation. The management plan should always include some tools for ensuring sustainability (e.g. reserves, barrier gear bans) and some other tools for raising revenues to pay for management (e.g. leasing, licensing). Management plans for some units may also include some tools for improving equity (e.g. bans of gears giving unfair advantages to their owners).

A wide range of different tools may be used to achieve each of these different objectives. Fishing activities, the environment and fish stocks can all, in some cases, be influenced by management. Some of the more useful management tools for floodplain river fisheries are listed in the following table, along with their primary objectives.

Where and when these tools are appropriate will depend not only on the technical features of the fishery (state of stocks etc.) but also on the capacity of stakeholders to carry out the

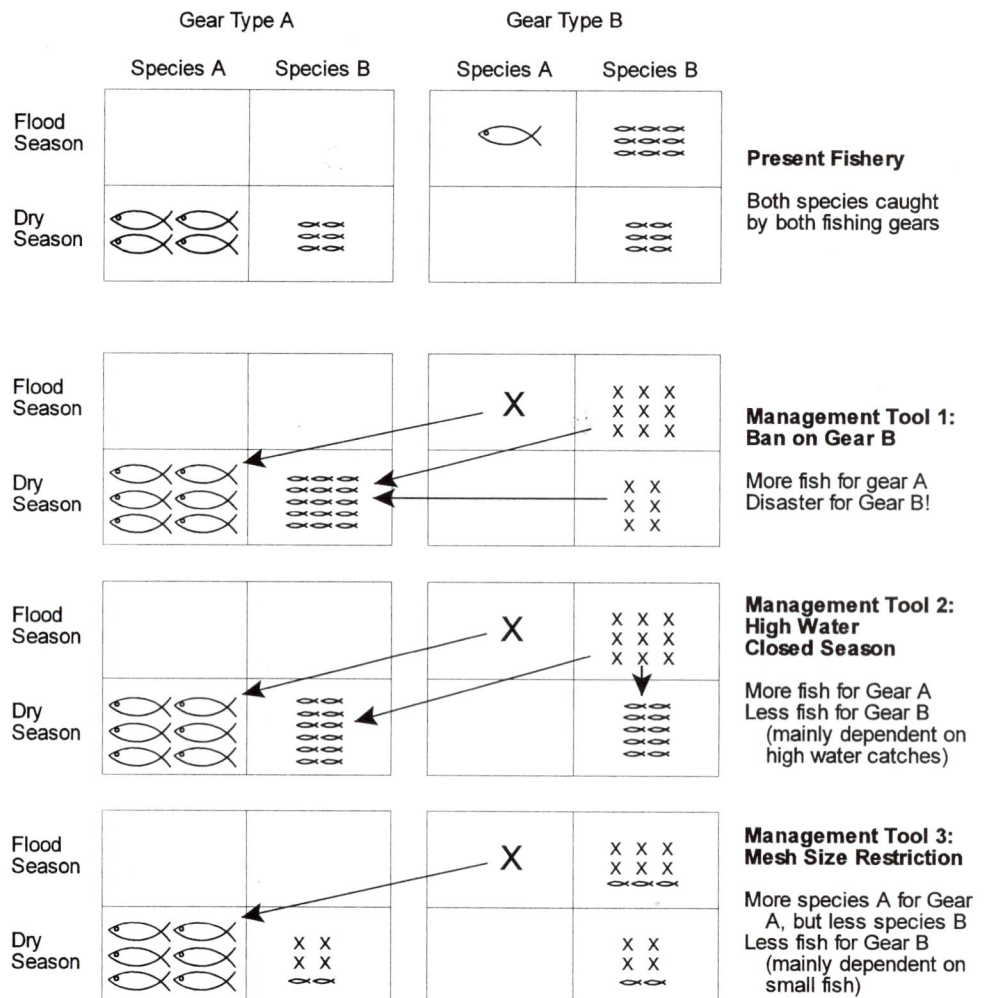

Figure 4.2
Hypothetical illustration of the effects of three management tools on the catches of two species, taken by two interacting fishing gears.

- *Gear A targets the larger Species A during the dry season, and takes a by-catch of the smaller Species B*

- *Gear B targets small fish including Species B and small specimens of Species A, but is restricted from the fishing grounds of Species A in the dry season*

- *Arrows indicate the shift in catches due to the management regulation*

- *Crosses indicate catches lost due to the management regulation*

Present Fishery

Both species caught by both fishing gears

Management Tool 1: Ban on Gear B

More fish for gear A Disaster for Gear B!

Management Tool 2: High Water Closed Season

More fish for Gear A Less fish for Gear B (mainly dependent on high water catches)

Management Tool 3: Mesh Size Restriction

More species A for Gear A, but less species B Less fish for Gear B (mainly dependent on small fish)

A menu of fishery management tools for floodplain rivers

Management Category	Management Tool	Primary Objective(s)
Managing the Environment	Environmental protection	Maintain overall integrity and productivity of river floodplain system
	Habitat restoration	Maintain primary habitats for fish spawning, feeding, and migrations
	Sluice gate management	Allow access of fish to polders (only in hydrologically modified floodplains)
	Water level manipulation	Maintain dry season water levels to maximise fish survival and fry production (only in hydrologically modified floodplains)
Managing *Who* Can Fish	Waterbody leasing	Raise revenue Reduce conflicts between fishers (control access)
	Gear licensing	Limit number of fishers / gears Raise revenue
Managing the *Amount* and *Type* of Fishing	Mesh / fish size limits	Limit capture of small / immature fish
	Reserves	Ensure some fish can survive the fishery to spawn and produce next year's stock
	Closed seasons	Limit capture of small / immature fish (flood season) Ensure some fish can survive the fishery to spawn (dry season)
	Dry season gear bans	Ensure some fish can survive the dry season to spawn and produce next year's stock
	Barrier gear bans	Allow access of (white)fish to spawning, feeding and survival grounds
Managing Fish	Once-only species introductions	Increase productivity of fish stocks, where appropriate species are missing
	Repeated fish stocking	Increase size of fish stocks, where natural breeding insufficient or depleted due to overfishing

roles needed to support the tools. In a newly established management area (village or catchment) it is important to begin with simple management interventions and allow co-operation and management capacity to develop. When a group of stakeholders have long experience of co-operation and have successfully managed their resource, the sophistication of technical interventions can increase.

Assuming that the group of stakeholders involved in VMAs will usually be easier to co-ordinate and manage than those in either IMAs or CMAs (bearing in mind the possibility of conflicts *within* communities), more detailed management interventions will usually be possible at the VMA level. Stakeholders managing IMAs and CMAs must be less ambitious and should focus on negotiating agreement between lower level units and on controlling permanent fishing structures and any especially threatening operations. A summary of the general suitability of these tools to the different management levels is provided in the following table.

The following sections describe in more detail how these tools may be applied in different management units. The accompanying tables should be read as menus of *potential* management options. At each level, the feasibility of these options will depend on the strengths and skills of the co-management partnership.

Management Category	Management Tool		VMA	IMA	CMA
Managing the environment	Environmental protection				✔✔✔
	Habitat restoration		✔✔✔	✔✔	✔
	Sluice gate management		✔✔	✔✔✔	
	Water level manipulation		✔✔	✔✔✔	✔
Managing *Who* Can Fish	Waterbody leasing		✔✔✔	✔✔	
	Gear licensing		✔✔✔	✔✔	
Managing the *Amount* and *Type* of Fishing	Mesh / fish size limits		✔✔	✔	✔
	Reserves	Whitefish		✔	✔✔
		Blackfish	✔✔✔	✔✔	
	Closed seasons		✔✔✔	✔	
	Dry season gear bans		✔✔✔	✔	✔
	Barrier gear bans		✔	✔✔	✔✔✔
Managing Fish	Once-only species introductions	Whitefish			✔✔
		Blackfish	✔✔	✔✔	
	Repeated fish stocking	Discrete waterbodies	✔✔	✔	
		Open floodplains		✔	✔

As mentioned earlier, the more complex management interventions such as fish stocking should be avoided until the managers are well accustomed to making collective decisions and enforcing them, and a commitment to common goals is well established.

4.3.1 Tools for Village (Blackfish) Management Areas (VMAs)

The following tools are suitable for use in VMA units to deliver benefits within the locality of the village. They are particularly designed for the protection of local blackfish stocks, but also to enable the maximum access of riverine whitefish into local fishing grounds.

A key requirement of VMA management plans must be the protection of local blackfish stocks throughout the dry season to ensure the success of spawning in the following year. Reserves, closed seasons and gear bans may best be used in *combination* to ensure some fish survival. Fishing communities will thus generally know where fish survive the dry season, and which gears particularly threaten them at this time. Management which prevents the use of such gears in *some* of the primary local waterbodies may effectively protect fish stocks without excessively restricting the fishery. Permanent, year-round reserves may not be necessary.

To implement such tools within a VMA unit, most villages will require considerable external support. Even if the village has an existing authority managing natural resources, new skills may be needed to take on responsibility for the fishery. The involvement of an NGO or external project, may be vital to the early success of a VMA. Government is

VMA management tools protect local blackfish stocks and ensure the access of whitefish to maximise local benefits

Management Tool	How (?), Objectives (→), Advantages (✔)	Disadvantages / Considerations
Habitat restoration	**?** Desilt blocked channels between river and floodplain waterbodies **→** Maintain access routes of migratory fish into village fishing grounds **✔** Village members may contribute labour	**✗** Labour intensive, depending on level of blockages
Waterbody leasing (may include sub-licensing of fishing gears)	**?** Divide village waters into discrete lease units **?** Allocate lease units by open auction (open to all bidders) to maximise revenue, or ... **?** Limit auction bidding to key stakeholders to improve co-operation with management or distribution of benefits **?** Use completely transparent lease system to prevent corruption **?** Use long term leases (>1 year), in most discrete waterbodies to encourage sustainable investment and management **?** Allow lessee to sub-license use of small fishing gears in lease unit **→** Raise revenues for village and management **→** Avoid conflicts of open access fishing **→** Encourage commitment to sustainable management **✔** Transfer temporary management responsibility to lessee (presence of lessee enables more effective enforcement) **✔** Enable lessees to gain some income from sub-licensees, but also limit their fishing to increase profitability of own large, more efficient but costly gears	**✗** Social exclusion of un-leased fishers **✗** Only applicable in floodplains subdivided by clear physical boundaries **✗** Knowledge that fish emigrate from lease-unit encourages high exploitation rates **✗** Short term leases may lead to over-exploitative extraction **✗** Long term leases may require credit facilities **✗** Long term leases may be resisted due to uncertainty of future profitability (so allow annual payment)
Fishing gear licensing (by village, in un-leased areas)	**?** Sell fishing licences for each gear type, specifying locality and season. **?** Where good fishing spaces are limited (e.g. for drift traps), use lottery to allocate licences **→** Limit overall fishing levels to ensure some fish survival **→** Avoid conflicts of open access fishing **→** Raise revenues for village and management **✔** Gear licensing may promote equitable access and high employment (useful where communities dislike exclusion created by leasing)	**✗** Social exclusion if privileged class created (especially if fishing positions determined by ancestral rights instead of lottery) **✗** May be difficult to determine the sustainable level of licensing (so use adaptive management) **✗** Difficult to monitor and enforce in larger management areas
Mesh / fish size limits	**?** Set minimum size limits for mesh sizes of each gear type and/or landing of fish **?** Select size limit to benefit most commercially important species in catch **→** Limit capture of small / immature fish **✔** Simple traditional concept **✔** May be enforced at local level by peer pressure or local agreement	**✗** Difficult to optimise in multi-species fishery, as gains for large species balanced by losses for small ones **✗** May exclude poorer groups from fishery (if traditionally based on capture of small fish) **✗** May require short-term sacrifice for long-term gain *Continued on page 38*

Management Tool	How (?), Objectives (→), Advantages (✔)	Disadvantages / Considerations
Dry season reserves / gear bans	? Restrict use of most dangerous dry-season gears (e.g. dewatering, poison, electric fishing, fish drives) in defined waterbodies ? Select deep, permanently flooded waterbodies ? Include some lake and some river waterbodies to protect different fish species → Ensure some blackfish survive the dry season fishery to produce next year's stock ✔ Easy concept, traditional and easy to formulate as law ✔ Visible and easy to enforce, especially if close to village, or in much-used waterway	✘ Limitation of fishing opportunities in reserved waterbodies ✘ Traditional users may resist bans on such highly effective gears ✘ May be unnecessary, if natural environment (tree snags, water depth) prevents dry season over-exploitation ✘ Reserves in large waterbodies may be difficult to enforce ✘ Enforcement of bans on small, portable, but effective gears (e.g. poisons) requires strong community support
Flood season closures	? Restrict fishing activities by all gears during early flood season → Limit capture of small / immature fish → Enable un-restricted migration of fish to spawning grounds → Ensure some blackfish can spawn without disturbance	✘ Exclusion of traditional users of flood season resources (often the poorer members of society ✘ Limits income and supply of fish from fishery to months outside closed season
Fish stocking	? Enclose waterbody (e.g. with barriers) to prevent escape of stocked fish ? Remove predators where possible to reduce fry mortality ? Stock valuable species of fish (usually every year) at start of flood season ? Combine with closed season / gear bans / size limits to limit capture of fish until marketable size ? Harvest fish just before or during dry season → Increase size of local fish stocks, where natural breeding levels insufficient → Increase value of catch available from fishery	✘ Costly ✘ Technical knowledge required for fry production ✘ Requires revision of traditional use and ownership patterns ✘ May exclude poorer groups from fishery (if fished during growth season, or using restricted gears) ✘ Requires infrastructure for funding ✘ Not generally good for biodiversity (does not matter on small scale but becomes significant on large scale) ✘ Requires raised level of education ✘ Requires repeat stocking every year
Barrier gear bans	? Restrict use of lateral barrier gears (between river and floodplain waterbodies), especially during early flood season → Enable access of whitefish into local fishing grounds → Enable un-restricted migration of local fish to their floodplain spawning grounds → Limit capture of small / immature fish	✘ May be difficult to monitor / enforce if barrier gears remain in place for later drawdown fishing season, but... ✘ Year round bans may be resisted due to loss of most valuable (drawdown) fishing opportunities

instrumental in increasing a VMA's capacity to manage by formally recognising the rights of these local stakeholders to manage the fishery through legislation. Government may also play a role in supporting the development of both technical and management skills within the village. As time passes and experience is gained, the reliance on external support should decline, and may even eventually disappear.

4.3.2 Tools for Catchment (Whitefish) Management Areas (CMAs)

At the catchment management level, government will have the greatest capacity to take the lead (see Section 3.5). Co-ordination and communication will be important, as other IMA and VMA management units will be nested within each CMA. CMA managers should look at individual

Tools for Catchment (Whitefish) Management Areas (CMAs)

Management Tool	How (?), Objectives (→), Advantages (✔)	Disadvantages / Considerations
Environ-mental protection	**?** Limit use of dams, channelisation and impoundments (if already exist, restore natural style of habitats where possible) **→** Maintain full range of habitats used by whitefish as breeding, nursery and feeding grounds **→** Maintain channels used by fish as migration routes	**✗** Costly, depending on scale **✗** May need research to establish appropriate habitats for restoration
Spawning reserves	**?** Locate in upstream whitefish spawning areas **→** Ensure some whitefish can spawn without disturbance	**✗** Exclusion of traditional users **✗** Difficult to enforce by government, unless supported by local users **✗** May need research to establish appropriate reserve areas
Barrier gear restrictions	**?** Restrict use of river-wide barrier traps, especially in main channels, and during flood-season, upstream migrations **→** Enable access of whitefish to spawning, feeding and survival grounds **✔** Easy for government to enforce (large, visible, stationary gears)	**✗** Exclusion of traditional users (barrier gears are efficient and may provide a large part of the local catches)
Fish species introduction	**?** Introduce high-productivity fish species into rivers where they are currently absent **→** Increase overall productivity of fish community **✔** Introduced fish may become self-reproducing (once-only activity)	**✗** Can only be decided at Government level and following specific international protocols **✗** Effects usually permanent and irreversible **✗** May have negative impact on biodiversity

VMA and IMA management plans in the context of the wider catchment issues and consult with lower units wherever there is conflict. The encouragement of VMA-level management may also itself be an effective means of improving the overall state of catchment resources. Illustrations of successful VMA management plans may be the best way of promoting further uptake of these ideas.

Decisions made in other sectors such as irrigation, infrastructure etc., which may adversely affect the fishery, are also often taken at the catchment level. Regional government offices are again in the best position to ensure fisheries arguments are heard. Governments may need support and training to establish good working relationships between their CMAs and the lower levels of

management units. NGOs and external projects may have important roles in this situation.

In addition to these co-ordination activities, CMA managers must also take the lead in the management of whitefish stocks for the overall benefit of the catchment. The tools listed in the following table may be most useful for this purpose. Barrier trap regulations and the general protection of the environment are thus used here at a catchment-wide level to ensure that the full migration pathways of whitefish are maintained. Due to their possibly negative *local* impact, such tools would not generally be proposed for local VMAs, and thus need to be supported top-down from the higher catchment authority.

Barrier trap regulations and habitat restorations are useful for CMAs to maintain the long-distance migration pathways of whitefish

4.3.3 Tools for Intermediate Management Areas (IMAs)

As discussed in Section 4.1.3, a range of 'IMA' floodplain systems fall in between CMAs and VMAs, being too small to qualify as full catchments, and yet too large to be managed independently by single villages.

Communication and co-ordination are vitally important roles for IMA units. External support from an NGO or project may be necessary to establish effective networks of information flow. Considerable care must be taken to ensure that institutional rules are fair, transparent and in line with the wishes of as many as possible of the various stakeholders. The relationship between the VMAs nested beneath an IMA or among the group of villages brought together under an IMA must also be clearly defined and widely agreed. This requires high levels of trust between stakeholders in neighbouring villages: it may take some time for a good working relationship to develop.

The following sub-sections illustrate tools which may be useful for IMAs working either as 'umbrella organisations' for several VMAs, or as a complete substitute for VMAs for large floodplain systems.

IMAs as umbrella organisations for VMAs

VMAs in isolation will not always be able to manage their local fishery effectively. If *blackfish* stocks overlap too much with those in other villages, a management decision made by one village and not the others will be less effective and may result in conflict. For example, gear restrictions or closed seasons made by village A, but not observed by village B, will benefit fishers in B, while fishers in A carry the cost. Hydrological modifications may also disrupt whitefish migrations (to the cost of all) or make some fish species highly vulnerable to overfishing by some villages (to the cost of others).

In these situations, IMAs may act as a forum for discussion between the different VMAs, and for their joint negotiations with the representatives of other sectors. While this may provide solutions, it will add an additional layer of complexity to the task of

management, requiring even greater co-operation and therefore support.

For IMAs covering a large lake, fish stocks are less likely to be vulnerable to fishing and environmental stresses in the dry season, so the primary management issues would be the control of overall fishing levels, the encouragement of co-operation, and the reduction of conflicts. For IMAs with large shared seasonal floodplains, the protection of dry season survival locations may be a more important issue. As with VMAs, the more complex tasks - particularly stocking - should not be attempted by IMAs without an appropriate combination of institutional maturity and outside support, probably from both NGOs and government.

Floodplain environments are often physically modified, or empoldered, to improve agricultural production. Sluice gates in such polders are usually used to modify water levels for the benefit of agricultural production. Within these areas, VMAs may be able to operate using the tools described in Section 4.3.1. In addition, IMAs may ensure that the interests of fishers are adequately reflected in decisions affecting the operation of the sluice gates. Where the polders are a physical barrier to fish movement on to the floodplain, sluice gates may be opened when possible to allow the entry of the maximum possible numbers of fish (eggs, fry and adults) into the impounded floodplain waters. Later in the season, sluice gates may be closed when possible to maintain water levels in the dry season waterbodies and increase the survival of fish. Such hydrological management may cause conflict between the agricultural and fishery interests so some IMA-level forum for resolving such conflicts will usually be needed.

IMAs as alternatives to VMAs

Where floodplain systems comprise many waterbodies spread over a wide area with no associations with any particular villages, they may be too large for the detailed management approaches used in VMA units. In such cases, IMAs may operate without the underlying structure and support of individual VMAs.

IMA Tools for Resolving VMA Conflicts and Enhancing Opportunities

Management Tool	How (?), Objectives (→), Advantages (✔)	Disadvantages / Considerations
Fishing gear licensing	? Sell fishing licences for each gear type, specifying locality and season. → Limit overall fishing levels to ensure some fish survival → Avoid conflicts of open access fishing → Raise revenues for village and management ✔ Gear licensing useful where habitat cannot be easily divided up into lease units	✘ Social exclusion if privileged class created ✘ May be difficult to determine the sustainable level of licensing (so use adaptive management) ✘ Difficult to monitor and enforce in larger management areas (need regular checking for licenses)
Fish stocking	? Enclose waterbody (e.g. with barriers) to prevent escape of stocked fish ? Stock valuable species of fish (usually every year) at start of flood season ? Combine with closed season / gear bans / size limits to limit capture of fish until marketable size ? Harvest fish just before or during dry season → Increase size of local fish stocks, where natural breeding levels insufficient → Increase value of catch available from fishery	✘ Costly ✘ Technical knowledge required for fry production ✘ Requires revision of traditional use and ownership patterns ✘ May exclude poorer groups from fishery (if fished during growth season, or using restricted gears) ✘ Requires infrastructure for funding ✘ Not generally good for biodiversity (does not matter on small scale but becomes significant on large scale) ✘ Requires raised level of education ✘ Requires repeat stocking every year
Reserves	? Restrict all fishing in defined areas ? Select deep, permanently flooded areas ? Select areas close to each village to spread access losses and management responsibility equally between villages → Limit total fishing pressure, to ensure some fish survive the fishery to produce next year's stock ✔ Easy concept, traditional and easy to formulate as law ✔ Visible and easy to enforce, especially if close to villages, or in much-used waterways	✘ Limitation of fishing opportunities in reserved areas ✘ Exclusion of traditional users, especially if poor or elderly traditionally fished close to village ✘ Requires good co-operation between villages for effective enforcement

IMA Tools for Managing Fisheries Inside Hydrologically Modified Areas

Management Tool	How (?), Objectives (→), Advantages (✔)	Disadvantages / Considerations
Sluice gate management	? Open sluice gates at time of maximum egg/fry densities outside gates ? Invite fisher's representative to advise sluice gate committees on fishery requirements ✔ Enable access of pre-spawning whitefish, and inflow of their eggs and larvae, in to polder to increase production	✘ Opportunities for opening gates limited by flood-water needs of agricultural sector ✘ Opportunities vary between years depending on rainfall patterns
Water level manipulation	? Close sluice gates during dry season to maintain water levels in polder ✔ Provide dry season habitats to increase blackfish survival and production of next year's stock	✘ Opportunities for controlling water levels limited by dry season needs of agricultural sector ✘ High dry season water levels may reduce effectiveness of some fishing gears (e.g. dewatering, fish drives)

In South Sumatra, Indonesia for example, the extensive River Lempuing 'lake district' has traditionally been managed by an open access leasing system where control of the waters is fully transferred to the leaseholders for the year. This solves the problem of managing distant waterbodies, but does little for the conservation of the resource (since the leaseholders attempt to take the maximum possible catch within the year of their lease). The following proposed modification of this system to *restrictive leasing* with some waterbodies withdrawn each year as reserves would maintain revenue generation from the resource, while also promoting conservation.

4.4 Adaptive management - monitoring for a reason

Floodplain fisheries managers have to deal continuously with uncertainty. As they operate in an environment that is both complex and variable, the results of their actions are difficult to predict and hard to evaluate. A management strategy based on repeated adaptation will assist in dealing with these uncertainties. Indications of when and where adaptation is needed will depend on a feedback or monitoring system.

4.4.1 Why monitor and adapt?

As noted in Section 4.2, it will usually be very difficult to predict the exact outcome of introducing a given management tool or institutional arrangement, due to the complexity of the systems affected and to local variations in habitat characteristics, social factors, external influences, and so on. As a result of these uncertainties, an *adaptive management* approach is recommended for managers at all levels of floodplain river fisheries.

Adaptive Management:

- explicitly recognises that the outcome of management actions can not be predicted;
- actively monitors and evaluates any management intervention or change;
- compares the outcome with that in other places or in previous times; and thus
- develops management strategies based on learning and feedback.

IMA Tools for the Management of Remote Waterbodies

Management Tool	How (?), Objectives (→), Advantages (✔)	Disadvantages / Considerations
Habitat restoration	**?** Desilt blocked channels between river and floodplain waterbodies **→** Maintain access routes of migratory fish into village fishing grounds **✔** Village members may contribute labour	**✗** Labour intensive, depending on level of blockages
Restrictive waterbody leasing / reserves	**?** Manage lease units as for VMAs, with associated objectives and advantages, but... **?** Remove some waterbodies from leasing system, as reserves (possibly rotating reserved waterbodies each year) **→** Ensure some fish survive the fishery (since total leasing may encourage overexploitation) **✔** Appropriate where no communities are particularly dependent on single waterbodies, or associated with them	**✗** Limitation of fishing opportunities **✗** Exclusion of traditional users? **✗** May be difficult to enforce reserves in very large systems, especially if in remote areas
Barrier gear bans	**?** Restrict use of lateral barrier gears (between river and floodplain waterbodies), especially during early flood season **→** Enable access of whitefish into local fishing grounds **→** Enable un-restricted migration of local fish to their floodplain spawning grounds	**✗** May be difficult to monitor / enforce if barrier gears remain in place for later drawdown fishing season, but... **✗** Year round bans may be resisted due to loss of most valuable (drawdown) fishing opportunities

The concept of adaptive management is illustrated in Figure 4.3. The processes of monitoring, evaluation and feedback are possible at various levels, both within individual strategies and for the management process overall. In either case the process is intended to increase knowledge of the effects of both technical interventions, such as gear restrictions, and institutional innovations. Ultimately this should improve management and so improve outcomes from the fishery.

Experimentation and learning can take place on a number of levels. At the VMA level, adaptive management is simply a process in which adjustments are made to the level of a management regulation or tool, or the mixture of tools being used, with the intention of improving the outcome from the fishery. If it is found, for example, that a new reserve does not increase the catch of 'species X' as much as hoped, it may be decided to introduce another reserve,

or to add a ban on a certain dry season gear for other nearby waters. Feed-back from monitoring data is thus required to detect whether or not these management strategies are working.

For managers at the IMA or CMA level the picture can be more complicated. On one hand, they may be trying to protect *whitefish* (e.g. through restrictions on barrier gears). Here they can follow an adaptive approach in exactly the same way as managers in VMAs. But they will also have a role in helping VMAs learn from each others experiences: comparisons can be made between the outcomes of the different management strategies adopted by different villages and news of successful procedures shared. One village may thus determine the most effective way to establish and manage reserves within its waterbodies, while another may identify improved ways to stock fish. Similarly, one village may develop a good

An adaptive management approach is needed for floodplain fisheries to fine-tune management tools (or select different tools) until the chosen objectives are achieved

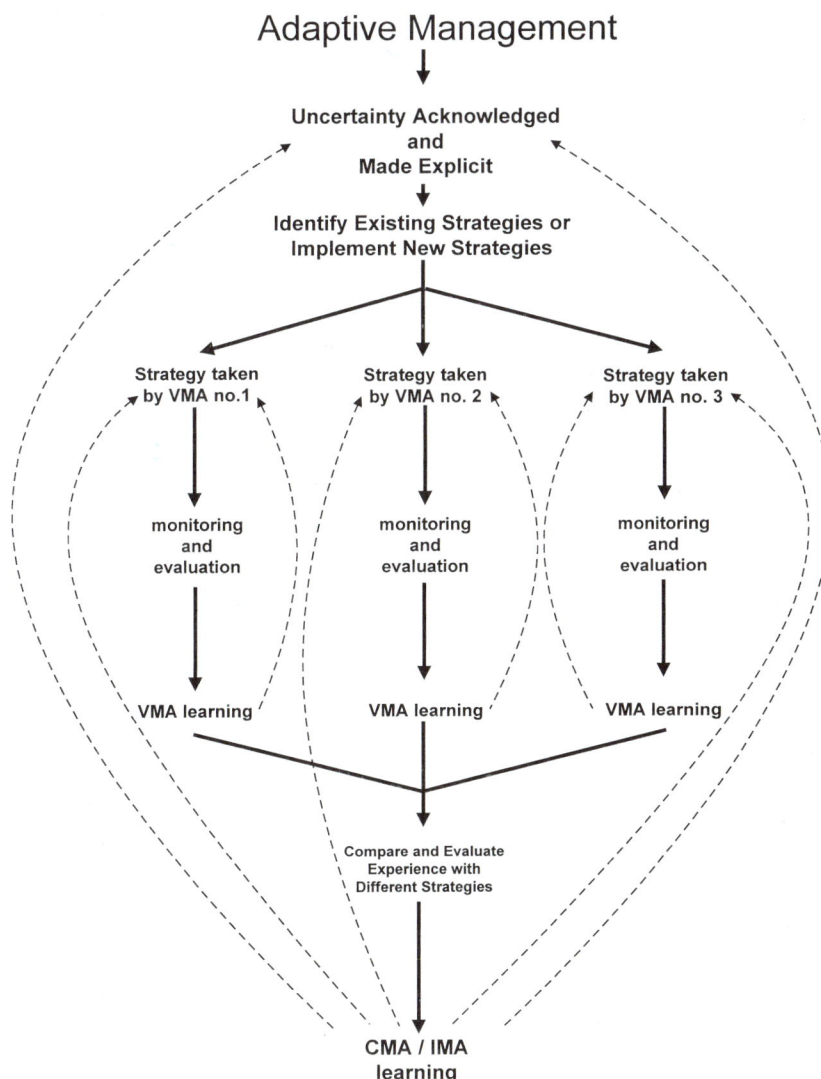

Figure 4.3 Diagram representing the possibilities for improved learning, between and within stakeholder groups, when adopting an adaptive approach to fisheries management

approach to resolving conflict while another has established a good system of communication within the village and with it's co-management partners.

How quickly lessons can be learned from adaptive management will depend on how actively adaptive strategies are implemented. Technically, there may be advantages in making large changes rather than small ones, since the results of small changes may be impossible to distinguish from normal variability in the environment. But large changes often carry higher risks, and taking big risks is not recommended, particularly in the early stages of establishing management capacity. As noted earlier, with improving management skills and confidence of the co-management team, the technical sophistication of the tools can increase and so too can the degree of risk the team is prepared to take. The temptation to attempt to impose strategies on VMAs to speed up the learning process should be avoided. Technical advice to VMAs must be directed to improving local livelihoods and reducing risk, not to increasing the knowledge at CMA or IMA levels.

4.4.2 Who should monitor?

As adaptive management of floodplain river fisheries takes place at all levels in the management hierarchy, it requires contributions to monitoring from all three major categories of stakeholder: communities, government and intermediary organisations.

Communities

Whether collecting monitoring data for VMAs or for CMAs, there will be real advantages in promoting the active participation of local fishers, including:

- fishers will be able to see, for themselves, the *impact* of the management strategy;
- fishers will be more likely to *believe* the data produced, if they are involved in its collection; and
- fishers may supplement the capacity of 'top-down' management agencies, who usually have insufficient resources and staff to monitor fisheries on their own.

In VMAs and the smaller IMA units, fishing community members have good

incentives to participate in the management and monitoring activities, as the results have the most direct relevance to themselves and their community. The degree of formality in this monitoring will, as mentioned above, depend on both stakeholder preferences and the type feed-back needed for higher levels.

Government managers should note that their community-based partners should not be expected to monitor themselves simply to relieve the work load of the fisheries department. Fishers may only contribute effectively to this process if they are actively involved both in choosing the management tools for their fishery, and the types of monitoring required for their adaptive management.

Government

Some fishing communities have traditionally practised 'adaptive' management approaches for centuries within their local area. In modern times, this process is becoming more difficult as local floodplain resources are increasingly impacted by the activities of a range of different sectors, sometimes well away from the local area. Such VMA-level managers may then find it difficult to determine whether changes in their fishery's outputs are due to their own management practices or to other influences from outside the village.

The adaptive management of a sub-divided catchment may therefore be most effective with both local monitoring *within* units such as VMAs, and comparisons *between* units at the CMA level. CMA-level managers may thus be most able to distinguish management impacts from those due to external factors by comparing VMAs with and without different types of management tools.

CMA managers are also best placed to monitor long-term trends in the catchment and take a broad perspective on many different activities within the catchment as a whole. Such a comparative adaptive management approach would also enable lessons to be learned in one village to be transferred to the local managers of other units. Though results will always depend on local circumstances, the effectiveness of management tools such as reserves, closed seasons and so on may then be determined gradually in each catchment, based on local experience.

Catchment managers, who will nearly always be government staff, should thus be responsible both for managing whitefish stocks at the CMA scale, and also for assisting the managers of all of the other smaller sub-units in the interpretation of their monitoring data. For the monitoring of whitefish stocks, CMA managers may need to rely both on data from government field staff, supplemented by data from any villages also managing their own fisheries for these species. While this 'adaptive co-management' strategy emphasises the decentralisation of management responsibilities to the lowest possible levels, it will still require major inputs from all levels of government.

Intermediary organisations

The role of NGOs or other intermediary organisations in monitoring will depend on their role within the management system. In some Asian countries, NGOs are important organisations in many grass-roots, community development activities. Where an NGO takes such a role within this system, their involvement in monitoring may include the organisation of participatory monitoring within the VMAs and the facilitation of flows of information between VMAs at the IMA/CMA level.

Intermediary organisations, such as research institutes or (other) NGOs, that are not involved in the day to day management can play a useful role in monitoring the operation of the institutional arrangements both at the VMA level and between the different partner organisations. Their independence as external agents may give them the objectivity required to analyse breakdowns and find solutions.

4.4.3 What to monitor?

Feed-back may take the form of an exchange of direct experience between fishers, a more elaborate internal data-gathering exercise within the community, or a formal monitoring exercise supported by outside agencies. The detailed *what* and *how* of monitoring is determined by the institutional context: who will use the data and how decisions will be made.

Detailed data collection by communities in *all* VMA-level units would be neither possible nor necessary in the heavily populated floodplains of Asia. Where VMAs are established and functioning effectively, formal data collection may be unnecessary for village-level decisions. Indeed, many traditional management systems have evolved successfully without formal monitoring. Management learning at VMA level can thus be based on 'common knowledge', derived from co-use of the resource in conditions where mutual observation is possible and secrets are hard to maintain. For management learning at higher levels, experiences might be passed to other VMAs by gatherings and exchanges or lessons distilled by outside field workers.

However there are a number of circumstances where a more formal monitoring system, that may include outside processing and analysis of data, will be appropriate. This may be where:

- new VMAs are being set up;
- the impact of untested initiatives at the village level need to be understood by NGOs; government agencies or donors;
- managers at the CMA/IMA level need to sub-sample VMAs to produce a quantitative evaluation of fishery performance; or,
- the status of *whitefish* stocks needs to be investigated.

In all cases, it must be remembered that formal monitoring data is expensive to collect and that evaluation is not always simple. Managers of higher level management units should therefore be sure that they need and will use the information that they request. This is particularly true where fishers are being asked to contribute data or bear the cost of its collection. Evaluation will be simpler for some types of data than others, and managers should be sure that they have the necessary analytical skills. Where evaluation involves *value judgements* of different stakeholders, it will be very helpful to decide *criteria* for evaluation before the outcomes are assessed.

Monitoring changes in the wider environment may help to explain unexpected changes in the fishery

The participation of fishers in monitoring programmes lets them see directly the impact that management is having

Catchment
managers
should help
local unit
managers to
understand
the changes in
their local
fisheries, and
pass on
important
lessons to
other units

The discussion below is directed mainly towards the collection of data at the more formal end of the monitoring spectrum. This emphasis reflects the interests of the likely readership of this document. It should not disguise the fact that, while the sorts of issues covered will be the same, most monitoring at the village level will tend to be informal.

In general, three different categories of monitoring data may be required. Firstly, information will usually be needed on the current state of the fishery, and the benefits being generated from it. Such data may relate to the abundance of fish stocks, the size of catches being taken, or the socio-economic benefits received by fishers. If these are increasing, it may indicate that the management measures are being effective. If they are not, additional data will be required to understand why.

Secondly, information on the *inputs* to the fishery is thus required, in order to explain the *outputs*. Influential inputs may include of technical and ecological factors, such as the amount of fishing and the condition of the river environment. Finally, the effectiveness of the management partnership and the management tools used will also be uncertain, benefit from adaptive improvement and therefore need monitoring. Guidance on the collection of the three types of monitoring information is given below.

4.4.4 Monitoring fishery benefits

For the more formal monitoring systems, the types of information which need to be collected depend on the objectives of management (see below). Since the achievement of nearly all these example objectives depends on the health of fish stocks, it will generally be useful to monitor the *ecological* state of the fish (e.g. their abundance). However it must be remembered that management interventions also affect other objectives and these must also be considered. For instance, the introduction of leasing, to increase government revenue, could reduce access by subsistence fishing households, thus affecting equity. Therefore, it is often be desirable to monitor the *socio-economic* state of the fishery (e.g. its current productivity, profitability and the distribution of benefits between stakeholders). To encourage village members to participate in management, it will always be useful to monitor at least some sort of indicator showing the social benefits being obtained from the fishery. The following sub-sections provide guidance on the collection of monitoring data on fish abundances and on socio-economic benefits.

Monitoring fish abundance - 'CPUE' data

If fish stocks are twice as abundant this year than last year, it may be expected that a standard unit of gear will catch twice as many of them. The *relative* abundance of fish may thus be estimated from '*catch-per-unit-effort*' (CPUE) data from the fishery. CPUE's may be estimated much more easily than the total catch of the fishery. They also indicate the current state of fish stocks more clearly than *total* catches, since the latter may kept high temporarily by increasing effort levels, even when stocks are in decline. CPUE figures should be estimated as follows:

$$\text{CPUE (kg / unit effort)} = \frac{\text{Total catch of species A, taken by gear B, in period C}}{\text{Fishing 'effort' of gear B, in period C}}$$

CPUE figures may be estimated either from a single catch of one fisher on one day ('period C' = 1 day), or from the combined total catch of several fishers over a defined period. In either case the measured catch should always be divided by the actual number of fishing effort units used in its capture. Where

If the objective is:	then, monitor:
revenue to government	total income from all leasing / licensing etc.
conservation of fish species X	abundance of fish species X
biodiversity of fish community	abundance of all individual fish species
profits of village members	income from fishing and associated costs
equity / benefit distribution	distribution of income / costs between stakeholders

several different CPUE estimates are available for a single gear type in a given period (e.g. from different fishers), an average CPUE figure may be calculated.

The *units* used for estimating CPUE are also critical, and vary between gear types. As shown below, the measure of *'fishing effort'* may need to indicate *how many* units of gear were used, their *size*, and *how long* they were fished for.

CPUE data should be collected to compare with figures from previous years, to illustrate whether fish abundance levels are rising or falling. When comparing this year's figure with earlier ones, managers should be aware that the *'catchability'* or effectiveness of fishing gears changes greatly between seasons. Catchability of barrier traps, for example, is highest when fish are migrating off the floodplains with the falling waters; catchability of many chasing and hoovering gears rises in the dry season when fish are most concentrated; catchability of set-and-wait gears may be highest when fish are actively foraging for food. *CPUE levels in the current year, must only therefore be compared with those for the equivalent periods in previous years.* Since the timing of the seasons varies between years, CPUE's may best be calculated as the average for each season (e.g. the wet season, the falling-water season and the dry season) rather than for individual calendar months.

Managers must also be sure that each gear type is only compared with the same gear type, and that the characteristics of that gear are not changing over time. If a new type of screen is used in a barrier trap, for example, its effectiveness may be increased and its CPUE's are then no longer comparable with those of the previous type. Those gears which are traditionally used in the fishery and are *not* likely to be modified in future offer the best prospects for monitoring fish abundances by CPUE levels. *CPUE's should never be averaged across different types of gears.*

If formal monitoring is being undertaken in VMA units, CPUE data may be recorded directly from fishers, perhaps by providing them with a simple form to complete on a daily or weekly basis. Managers should realise that some fishers may not wish to complete such forms, for a variety of reasons. Managers should therefore be prepared to offer full explanations of the adaptive management strategy. Providing fishers with cash or other incentives for their co-operation may or may not be appropriate. In licensed or leased fisheries, the management agency may be able to insist that catch and effort data is submitted by licensees / lessees, as a condition for their access to the fishery or the auction next year.

In CMA-level management units, agency staff may need to collect their own CPUE data. In this case, staff should only sample data direct from fishers, at wholesale (first-sale) markets, and *not* from traders at public (secondary) markets. At the market, samplers must be sure that the catch they are measuring

CPUE data may be compared with figures from previous years to illustrate whether fish abundance levels are rising or falling

Measures of fishing effort for different gear types

Gear Type	Measure of fishing effort
Set-and-wait gears	Number of gear units (standard-sized gear, set for a standard period, e.g. per trap, when always set overnight) Number of gear units x time set (standard-sized gear, set for a variable period, e.g. per trap per hour) Number of gear units x size x time set (variable sized gear set for a variable time period, e.g. per *metre* of gill net per hour
Chasing gears	Number of gear units x time actively fished
Barrier gears	Number of barriers trap units x time set
Hoovering gears	Usually for a stated waterbody, with no specific units. Assuming standard fishing practices, CPUE's may be simply expressed as the total catch from the waterbody by that gear type. The success (catchability) of hoovering gears, such as dewatering, may however be strongly affected by dry season water levels: this may make CPUE's difficult to compare.

Socio-economic
monitoring
programmes may
demonstrate both
the overall profits
of fishing, and
their distribution
between different
stakeholders

represents the *whole* catch of only one gear type, and not a mixed catch from several gears. They must also determine the amount of fishing effort directly from the fisher, since this will rarely be known by fish traders or anyone else.

Monitoring socio-economic benefits

Evaluation of the contribution of management measures to increasing income or improving equity requires an understanding of both the net value of catch and its distribution between different stakeholder groups. Relying on data collected by routine monitoring for such an analysis is not recommended. If there are additional social objectives such as empowerment of the poor, these will also be difficult to evaluate through simple data collection techniques.

In formal monitoring systems, changes in the economics of the fishery can be monitored by collecting data on both the *value* of catches and the *costs* of fishing. Fishing costs may need to be subdivided between the costs of time, fishing gears, bait and so on, and the fees paid for access to fishing grounds, e.g. by sub-licensees. In many of the more open floodplain fisheries, great care is needed in extrapolating the results from individual households.

In addition to seeing whether socio-economic benefits are rising or falling over time, monitoring may be required on the *distribution* of impacts on different stakeholder groups, and their sub-categories:

* fishers of different types (professional, subsistence etc.);
* leaseholders;
* other community members (e.g. as lease payments paid into a village development fund);
* the fisheries department; and,
* local and national government administration.

Estimation of this distribution will usually require more detailed survey designs than needed for simple CPUE or income data, and will always need to be adapted to local conditions and fishery structures. Information will usually be required on the total numbers of fishers in different categories, the gears they

use, and the average catch values and fishing costs for each gear type. This information is often highly sensitive and fishers may be particularly reluctant to declare it. Managers may need to guarantee the confidentiality of the information to protect the interests of the fishers. Presentations of socio-economic monitoring data may need to be averaged to avoid revealing sensitive information about individual fishers. Wherever possible, however, the monitoring and reporting process should be made as '*transparent*' as possible, to minimise the submission of false data.

4.4.5 Monitoring inputs to the fishery

As well as monitoring the *outputs* from the fishery (as related to the objectives), managers may also need to monitor any changes in the *inputs* to the fishery. In addition to the application of the management tools, important factors may include both the overall level of fishing (the numbers of fishers and gears etc.), and environmental factors such as water levels, and changes in land use patterns. Such data may help to explain observed changes in the fishery which do not seem to be due to the changes made to the management strategies and technical tools.

Monitoring environmental conditions

Though over-fishing may change the species composition of the multi-species catch, it rarely reduces the *overall* catch from the fishery. Total catches may, however, be reduced by changes in the environmental quality of the floodplain system and the wider catchment. Impacts on water quality are particularly important here, as are modifications of critical fish habitats (e.g. spawning habitats) by users of other sectors.

In addition to man-made impacts, environmental conditions also change *naturally* from year to year, causing significant variations in the productivity of fish stocks. Managers should thus expect their fish catches (and other monitoring outputs) to change significantly between years. Management actions should be made on the basis of long-term *trends* in outputs, rather than a simple

Since
environmental
conditions
change naturally
from year to
year, managers
will need to
examine long-
term trends and
not expect
immediate
answers

comparison between this year and last year. Environmental data may also be collected to help interpret year-to-year variations in outputs.

Important environmental factors to monitor thus include:

- changes in nearby land use patterns, particularly where natural floodplain habitats are converted for alternative uses,
- increases in pollution inputs, either from agricultural pesticide use, industry or urban settlements,
- water levels (the height and duration of the flood determines the productivity of fish stocks; and the length and dryness of the dry season determines the effectiveness of dry season fishing gears and the resulting survival of the spawning stocks).

Monitoring the amount of fishing

High levels of fishing may be responsible for declines in preferred species, and may *not* be prevented by some types of management tools. Having a reserve, for example, may not prevent the build up of very high levels of fishing in the waters around it. Monitoring the *overall* amount of fishing in terms of the 'fishing effort units' discussed previously is difficult for a multi-gear fishery, due to the differences between effort units. Managers should instead aim to collect simpler comparative data on fishing, such as:

- the overall numbers of fishers working within the unit;
- the numbers of different types of fishing gear in use (particularly those gears which catch species which seem to be declining);
- the introduction of new fishing gears, or improvements in the effectiveness of old gears; and/or,
- known or suspected levels of poaching (especially of fishing in reserves, which may then *not* conserve fish stocks as much as intended).

4.4.6 Monitoring the change in management approach

The adoption of co-management approaches is a significant change in

the way floodplain fisheries are managed and will take some time to develop. The institutional arrangements to support this development are a critical influence on the effectiveness of the approach in meeting its objectives. Developing a set of arrangements that are, and remain, appropriate is therefore an important task.

In keeping with the cyclical, adaptive approach to the technical aspects of fisheries management discussed above, the approach to developing this new style of management should avoid rigid predetermined formats. Rather, it should attempt to develop solutions through continuous self-examination and repeatedly updating and improving its approach. This requires monitoring of the stakeholder groups in a co-management agreement and the way they co-operate in order to manage.

Development as a process

Recent literature has placed particular emphasis on the idea of development as a *process*[1]. This emphasises the importance of retaining flexibility, learning from experience, and recognising both the importance of social context of outcomes and the dynamic nature of development. This contrasts to the more conventional view of making such changes using a very controlled or 'blueprint approach' with a fixed and predictable relationship between project inputs and project outputs.

Traditional monitoring systems are of limited use when the development of a new approach is treated as a process. These monitoring systems are mostly based on a set of predetermined indicators, intended to quantify particular outcomes: they will rarely show *why* such outcomes are achieved. Monitoring of *organisational* impacts, such as enhanced capacities of stakeholders, changed perspectives or empowerment, is particularly difficult. New kinds of information generation and communication are required. In contrast to more traditional projects (such as road construction or agricultural input delivery),

[1] See for example: Mosse, D., J. Farrington and A. Rew (Eds.), 1998. Development as Process: Concepts and methods for working with complexity. 1st ed. Routledge, London. 202 pages.

Adopting co-management partnerships is a significant change in management of floodplain fisheries: difficulties should not be underestimated

An openness to learning will help develop this new approach to management

Solutions will be found by continual self examination and repeated improvement of approaches

development of a new system of fisheries management can not be judged from a single objective or one professional perspective. If obstacles to an effective co-management partnership are to be resolved, the perspective of *each* stakeholder group must be understood and taken into account.

Monitoring of the effectiveness of the change to a co-management style for floodplain fisheries must therefore examine those factors which determine both the effectiveness of individual organisations and the effectiveness of the relationships between them. The success of these relationships will influence all aspects of the creation, adoption and enforcement of management rules and the resulting outcomes from a fishery.

The explicit use of process-based approaches in fisheries management has been very limited, but guidelines can be obtained from experience in other sectors in Asia. For example, as co-management partnerships are being established, monitoring the process of their development should be carried out on a continuous basis. This will pick up successes and failures early and allow improvements to practice to be made before styles of management become too established and difficult to change. When the co-management partnership has become more experienced in its relationships, process monitoring may become less frequent, perhaps with shorter, more issue-focused visits, or when support is requested by the partnership.

Another useful area of guidance from work in other sectors refers to who should be involved in process monitoring. Breakdowns in the co-management partnerships will often be due to conflicts between the main partners or result from inadequate capacity of a partner to carry out their agreed roles and responsibilities. Therefore, it may be helpful to include researchers from outside the core partnership to help with process monitoring. The intermediary organisations such as research institutes, other NGOs or projects may have critical inputs here. Such contributors should ensure, however, that their own perspectives and beliefs

do not influence their judgment of unfamiliar local situations.

Though invaluable in providing insights about the management partnership, process monitoring will not be easy. Difficulties may arise if managers feel threatened by researchers who document unplanned as well planned outcomes, especially if the local managers feel criticised by the outsiders. The modifications to improve the process of changing management, which are the objective of such process monitoring, may also be difficult to achieve if they are seen as signals of failure by outside funding bodies. Both of these problems can be moderated if the approach is explicitly adaptive, and the need for periodic change is seen as being normal rather than as a failure to control resources effectively. *Process monitoring should be a means of developing stakeholders' capacity for participation, and not as a means of allocating blame for management failure.*

Though process monitoring is now being used in the management of natural resources, it may still be unfamiliar to many fisheries departments. Much further work is required on its use for the development of fisheries co-management partnerships. In the mean time, fishery managers should recognise the need for this type of monitoring, and request advice from wherever it may be available, e.g. other government departments where it has been used, NGOs or international development projects.

4.4.7 Paying for monitoring programmes

The collection, analysis and presentation of data from monitoring programmes is essential for effective management, but requires significant inputs of both time and money. The costs associated with such activities should ideally be raised from the fishery by some component of the management plan, such as a leasing or licensing scheme. Authorities who receive money from such existing schemes should also be aware that some of these funds may need to be reallocated to pay for the improved management of the fishery.

The funding required for monitoring programmes should be raised from the fishery

5 Steps to Successful Management

This section summarises how the ideas in the *what, who* and *how* sections may be drawn together into a successful management strategy for floodplain river fisheries. The various steps are divided into those which should be taken by (1) *national*-level policy makers, (2) *catchment* managers (including any co-ordinating fora for multi-CMA rivers), and (3) managers of the individual CMAs, IMAs and VMAs. The steps are summarised in Figure 5.1 below and described more fully in the following tables.

Many of the activities described in these tables will be new to governments, or involve new forms of partnership with other organisations. It must therefore be emphasised that adoption of the proposed strategy should be built up slowly, using a 'process' approach, and by learning from successes and failures.

Institutional arrangements - whether local fisheries management groups or higher level partnerships - are critical to the success of these approaches. These can not be defined in advance or imposed fully formed from above; they must also be allowed to evolve and develop, with effective participation from below. Time must be allowed for differences to be negotiated and conflicts to be resolved. Again, the approach taken should be gradual and should focus on learning continuously from experience, rather than on promoting a blueprint approach.

A useful first stage should be for fishery departments to develop a partnership with organisations experienced in facilitating the development of community organisations. Village management units should then be promoted, initially in simpler situations

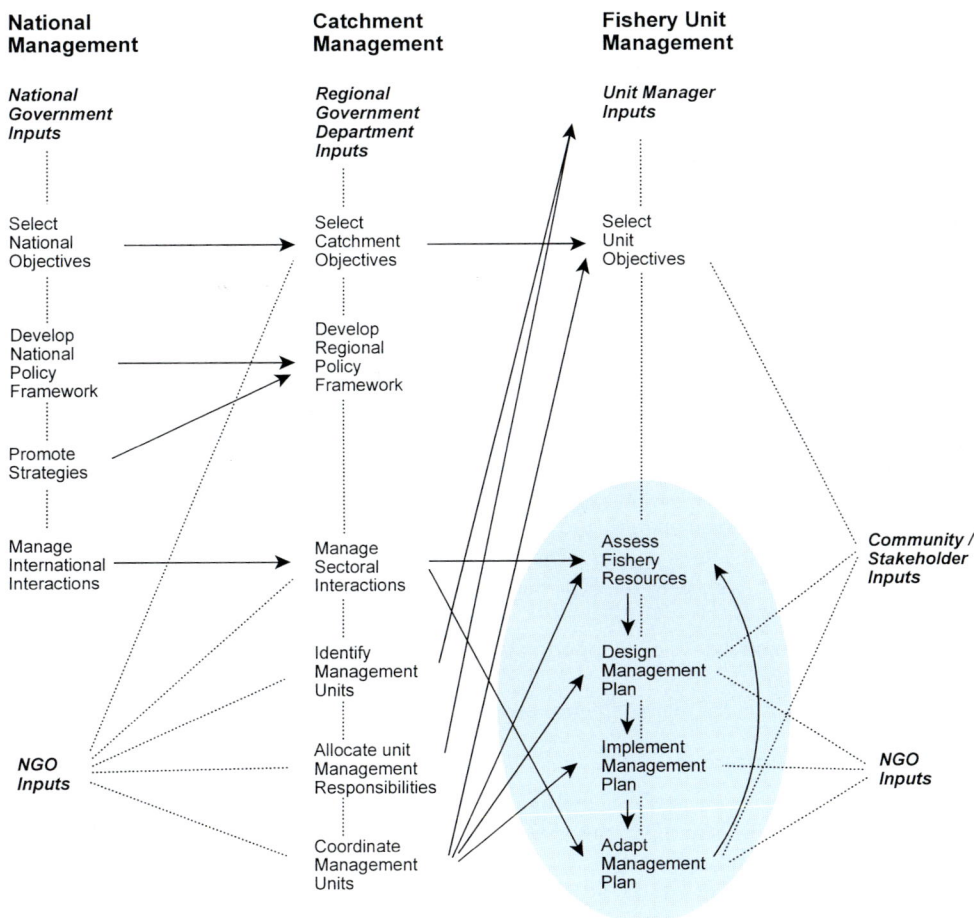

Figure 5.1 Management activities required for effective co-management of fishery units (dotted lines indicate inputs by different organisations; solid arrows indicate the contribution of the information between different activities; the shaded area indicates the main components of the ongoing adaptive management process)

where waterbody control is relatively undisputed or where traditional institutions already exist. Activities for resolving conflicts between VMAs, or for developing IMA-level management should come later.

5.1 National level (leadership, endorsement and legitimisation)

The main responsibilities of national level fishery managers (the fisheries ministries / directorate generals etc) is for the promotion of improved management systems, and the endorsement of activities at the lower management levels. Decentralised management can not proceed effectively until the rights of local people and agencies to manage is recognised and clearly stated in the legislation.

5.2 Catchment level (regional leadership and co-ordination)

Management activities at the catchment level provide the necessary leadership and co-ordination of the lower VMA and IMA management units (see table). Managers at the catchment level must also be responsible for the management of whitefish stocks in the CMA-level management units (see Section 5.3).

Catchment-level management activities may be undertaken by any appropriate administrative level below national government. Some countries may have two or even three administrative levels which could each participate in these management activities at appropriate stages. Where spatial administrative units do not overlap exactly with river catchments (as will often be the case), catchment level management may need to involve collaboration between two or more administrative regions. Such collaboration may either involve the creation of a new catchment management forum, or the writing of a memorandum or understanding between the existing units.

5.3 Management unit level (management of fishery resources)

The management activities in the final table provide for the sustainable, long-term management of the fisheries in each management unit. They should be undertaken by the co-management partners of each CMA, IMA and VMA unit, according to their interests and capacities.

National-level management activities

Select national objectives (Government)	• Specify broad objectives (e.g. sustainable resource use, support of community objectives) rather than detailed targets, allowing finer specification at lower management levels. • Ensure compatibility with existing national plans and international agreements for conservation targets (e.g. UN Convention on Biodiversity, Agenda 21).
Develop national policy framework (Government)	• Develop legislation which enables and supports decentralised co-management. • Enable adaptive management approaches by allowing rapid creation and adjustment of local management rules, with the minimum possible formal legislation.
Promote regional adoption of co-management strategies (Government)	• Provide training and extension, supported by appropriate materials (e.g. translated copies of these guidelines). • Communicate lessons of management between catchment managers.
Manage interactions for international rivers (Government)	• Discuss water resources and fisheries sharing for international rivers with national-level managers of adjacent countries.

Catchment-level management activities

Select catchment objectives (Government Fisheries and Regional Planning Departments)	• Ensure adaptation to local conditions and requirements, and compatibility with national-level objectives. • Ensure the maximum possible agreement with the interests of all stakeholders, but... • ... select *compatible* objectives (not all objectives can be achieved at the same time).
Develop regional policy framework (Government Fisheries and Legal Departments)	• Adopt modified national legislation enabling decentralised co-management and rapid rule-making for adaptive management by fishery units.
Manage sectoral interactions (Government Natural Resources Departments)	• Identify natural resource exploitation zones for shared or exclusive use by fishers, agriculture, industry, national parks etc. • Investigate negative impacts between sectors, and promote understanding or mitigation. • Communicate the anticipated or estimated impacts of each sector on the productivity of local fisheries, to local management units to assist their adaptive management.
Identify management units (Government and Unit Managers)	• Identify suitable CMAs, VMAs, and IMAs, based on spatial distribution of floodplain systems (hydrology), migrations of fish stocks (blackfish and whitefish), and distributions of fishing communities. Use the interview checklist in all villages close to waterbodies, to provide the necessary understanding of the catchment's resources. • Build on any traditional management institutions still used successfully by fishing communities or local administrations
Allocate unit management responsibilities (Government and Unit Managers)	• Allocate management responsibilities to co-management partners where the listed 'conditions for co-management success' are met (see Section 3.5).
Coordinate management units (Government and Unit Managers)	• Ensure the compatibility of objectives for 'nested' units (e.g. VMAs within CMAs). • Promote the collaboration of management activities between adjacent units, particularly to enable migration of whitefish stocks. • Judge and resolve conflicts between units, as required. • Train unit managers, as required. • Provide catchment perspective for interpretation of local monitoring data • Communicate lessons of management between fishery units.

Management unit-level activities

Select unit objectives (Unit Managers and other Stakeholders)	• Hold meeting(s) with unit stakeholders to discuss local fishery resources, and agree the local objectives for management of the fishery unit. Publicise the agreed objectives. • Ensure selected objectives are well adapted to local conditions and requirements, and compatible with national and catchment-level objectives. • Ensure the maximum possible agreement with the interests of all stakeholders, but... • ... select *compatible* objectives (not all objectives can be achieved at the same time).
Assess the fishery resource (Unit Managers)	• Interview fishers to measure the current outputs from the fishery, and compare with the selected objectives (are the objectives already being achieved, or is restrictive management required?). • Interview fishers or sample present catches to determine the relative importance of local blackfish and migratory whitefish in the catches of the unit. • Interview fishers about any declines in fish stocks, or historical changes in the species composition of fish catches (expect the large, most valuable fish species to decline first). • Interview fishers (see checklist) to determine the interactions between fishing gears in the fishery, and the likely impacts (both positive and negative) of different management strategies. • Review any scientific literature available for the fishery (from Fisheries Department?).

Design the management plan (Unit Managers and other Stakeholders)	• Hold meeting(s) with stakeholders to discuss the results of the fishery assessments, and design an appropriate management plan to achieve local objectives. • Design an integrated 'technical strategy' (a combination of different management tools), to achieve (1) conservation of fish, (2) raising of funds for management, and (3) fair distribution of social benefits from the fishery. Select from the tools listed for each type of management unit (Section 4.3). • Design a complementary 'institutional strategy', indicating *who* will be responsible for each part of the management plan (rule setting, monitoring and enforcement, revenue management, future assessments, etc.) • Ensure that the plan has the greatest possible local benefits, and the minimum possible negative impacts (ensuring its support by local people, and reducing the requirements for strong enforcement). Discuss giving compensation to badly affected fishers.
Implement the management plan (Unit Managers)	• Enact local legislation for management tools, if required. • Publicise the agreed management plan at public meetings, and with announcements on village notice boards etc., stating any new rules clearly and precisely (what are the new rules?; what are they intended to achieve?; when will they start?; what are the penalties for offenders?). • Enforce management rules, penalise offenders as required. • Resolve conflicts between fishers within the unit, and with outsiders. • Communicate monitoring data to catchment-level managers.
Adapt the management plan (Unit Managers)	• Monitor fish abundances (CPUE's) and socio-economic outputs from the fishery, as required to determine whether selected management objectives are being achieved. • Monitor the amount and type of fishing (numbers of fishers, use of different fishing gears, introduction of new gears, prevalence of illegal fishing etc.), to determine whether the selected management tools are limiting fishing as intended. • Monitor local environmental conditions (water levels etc.), to determine their likely impact on the current productivity of the fishery. • Be aware of likely impacts of developments in other sectors within the catchment, such as loss of floodplain habitat to agricultural use, or loss of upstream fish spawning grounds (information to be provided by catchment managers). • Each year, compare the current outputs from the fishery, with those in previous years, particularly looking for *trends* in the outputs. Jointly examine the current fishing practices, local environment conditions, and wider catchment influences to determine what may be responsible for current output levels. • If the fishery outputs are not meeting the unit objectives due to continued high levels of fishing, or to changes in fishing practices, adjust the technical management strategy as appropriate (change the *level* of regulations, e.g. the length of the closed season, or add *new* management tools). • If the fishery outputs appear to be declining due to the impacts of other sectors, communicate results to catchment managers for discussions between sectors.

Bangladeshi floodplains support enormous numbers of fishers - some people own no fishing gears but may still catch a few small fish by rolling up a circle of weed (All photographs by D.D. Hoggarth)

The Komering River in Indonesia now dries up almost completely in the dry season due to irrigation works upstream

Bangladeshi fishers must operate alongside farmers who use the floodplains to grow rice for much of the year

Large-scale flood control schemes may give agricultural and social benefits, and create some fishing opportunities for 'jump trap' fishers, but they may also have negative impacts on the other fishers inside the scheme

Bangladesh's heavily exploited floodplain waters support millions of fishers, but catches now mainly comprise the very smallest fish and prawns

Fine-meshed fish traps usually indicate the most heavily exploited fisheries

Barrier traps such as corong flume traps and tuguk suspended trawls in Indonesia are expensive to construct, but highly effective at catching migrating fish

In Bangladesh, dry season floodplain waterbodies may even be pumped dry to extract the last remaining fish

Detailed scientific studies may provide a good basis for understanding aspects of the fishery but are only a small part of the overall management strategy

Rights to fish in some Indonesian waters are auctioned every year - such existing links between government and fishers provide good opportunities for developing further collaboration

Cheap 'set-and-wait' gears such as long-lines provide an income for the poorer fishers but may be negatively affected by regulations intended to control other gears

Even small floodplain waterbodies may enable many brood fish to survive the dry season, if protected as reserves (River Ganges, India)

Main river channels, such as the River Ganges in India are too big for the effective use of barrier or hoovering fishing gears and provide natural refuges for fish

Small channels draining the floodplain are used by migrating whitefish, but may be easily blocked by barrier traps (River Lempuing, Indonesia)

Sluice gates which provide fishing opportunities for a few 'jump net' fishers may also be operated to maximise fish migrations for wider benefits to the fishing community

Changes in the inputs to the fishery, such as the gear types in use, should be monitored in order to understand any changes in the monitored outputs